Inside **MATHEMATICS**

Geometry

UNDERSTANDING SHAPES AND SIZES

Inside **MATHEMATICS**

Geometry

UNDERSTANDING SHAPES AND SIZES

(x,y)

Mike Goldsmith

SERIES EDITOR: TOM JACKSON

𝓍

SHELTER HARBOR PRESS
NEW YORK

Introduction

Geometry is the study of shapes and of spaces. The word means "land measurement," because it was first used by farmers and builders, and still is today. However, it is much more than that, and allows us to study places we cannot go or even imagine, like the beginning of the Universe, the far reaches of space, and other dimensions.

Practical math

Geometry existed long before even the simplest equations did, and for thousands of years it was almost the only kind of mathematics there was. For instance, while today we work out square roots using an equation (or an app, calculator, or computer spreadsheet with the equation built in), a Greek student 2,000 years ago would use the geometry of a half circle, as shown below:

Geometry is based on methods used to measure land, but it soon became a way of performing calculations.

To find the square root of a number (represented here by the length x), add 1 to it and draw a line of this length (1 + x). Use this line as the diameter of a circle, and then draw a vertical line from a point which is a distance x from one end of that line to the circle. The length of this line is the square root of x. An ancient Greek would have left the answer as a length, not as a number.

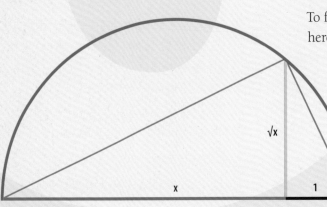

√x

x

1

All triangles have angles that add up to 180°. This was discovered by the ancient Greeks, though they did not use degrees. Instead, they said that "the angles in a triangle add up to two right angles" (a right angle is 90°).

90°

50°

40°

60°

60°

60°

77.5°   77.5°

25°

60°

60°   60°

60°

30°

90°

Real space

Geometry has also long been used to study the Universe, telling our ancestors the shape of the Earth, the size of the Moon, and the distance to the Sun. Until the 19th century, geometers assumed that the triangles and squares they used were exactly like real ones. But then some adventurous mathematicians began to explore how geometry might work if nature used a different system. For instance, according to the geometry developed by the ancient Greeks, every triangle has angles that add up to 180°, and an area (A) equal to half of the triangle's base length (b) multiplied by its height (h): A = 1/2b x h.

Other geometries

Nevertheless it is perfectly possible to change the rules of geometry and study other kinds of triangles, with angles that add up to more or less than 180°, or even triangles whose angles change according to its size. Using a kind of geometry with shapes like that might mean that a plan of

a building or city would be useless—unless of course the plan was the same size as the building or city itself! These new kinds of geometry are called non-Euclidean, to distinguish them from the geometry developed by Euclid, a Greek

About 2,300 years ago, Euclid wrote a book about geometry called *Elements*. It became the single most important mathematics book in history.

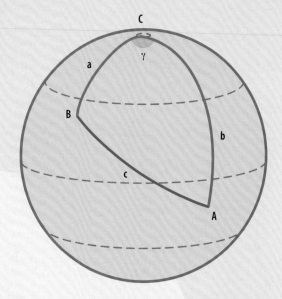

A triangle connecting three points A, B, and C on a globe has the sides a, b, and c. The sides appear straight when measured on the surface, but the triangle they create has angles which add up to more than 180°.

mathematician who was the greatest geometer of the ancient world.

Relativity

Until the early 20th century, non-Euclidean geometries were not of much use, but then physicist Albert Einstein made a breakthrough. He found that non-Euclidean geometry could be used to explain the nature of gravity. Amazingly, careful measurements based on his work revealed that the geometry of the Universe really is non-Euclidean. For instance, surveyors and scientists use laser beams to make measurements because

they travel in straight lines. However, if you sent laser beams between Mercury, Venus, and Earth, you would get a triangle whose angles would add up to slightly more than 180°.

A black hole warps the space around it, changing space's geometry so that straight lines are curved.

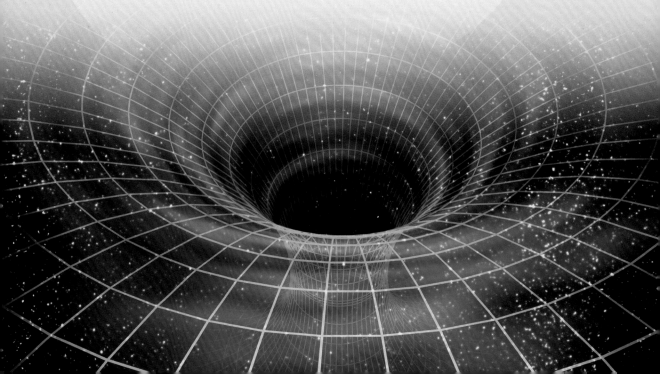

Squares and cubes

Another geometrical idea long thought to be imaginary is the fourth (and even higher) dimensions of space. Lines are one-dimensional, flat shapes are two-dimensional, and solid objects are three-dimensional. Simple math links lines, shapes, and solids. For example, the area of a square with edges of length L is $L \times L = L^2$ — that's pronounced as "L squared." The volume of a cube with edges of length L is $L \times L \times L = L^3$ ("L cubed"). The ideas of squaring and cubing are useful quite apart from their meaning in geometry. For instance, in a car crash, the force of the impact increases with the square of the car's speed.

Going further

There are also some relationships that depend on multiplying something by itself four times. For example, when a liquid flows through a pipe, the ease with which it flows depends on the pipe's width (W) multiplied by itself four times, or "W x W x W x W." We can just abbreviate this to "W^4," which is read out as "width to the fourth power" or just "width to the fourth." And we can extend this as much as we like. For instance, 3^{10} ("three to the tenth") means $3 \times 3 \times 3 \times 3 \times 3 \times 3 \times 3 \times 3 \times 3 \times 3 = 59{,}049$. A few people in the

Albert Einstein's famous theories were based on the geometry of space and time.

19th century did wonder if there might really be a fourth dimension of space. They would have been astonished to hear about the latest scientific theory of the whole Universe, string theory. String theory may not be correct, but if it is then that means that there really are very tiny objects with not just three, or four, but ten dimensions! Geometry, the most ancient form of mathematics, definitely has a future.

A line has one dimension, and if we add lines to it at right angles we get a two-dimensional shape, such as a square. Adding more squares at right angles makes a cube, which is three-dimensional. It seems impossible to take this process any further to make a four-dimensional object, yet mathematicians believe that such objects exist.

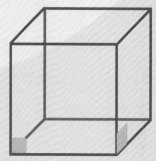

The Roots of Geometry

ALL ANCIENT CIVILIZATIONS USED MATH, BUT GEOMETRY REALLY BEGAN WITH THE NILE, THE GREAT RIVER WHICH RUNS THROUGH EGYPT. The population depended on the Nile for their livelihoods, but the river was not the easiest of neighbors. Every year its floodwaters would spread across the land, washing away soil and plants, and ruining many of the small food plots, even as it deposited new fertile mud.

Egyptian farmers depended on the Nile floodwaters which would spread across the surrounding lands.

All riverside land belonged to the pharaoh. King Sesostris III (who ruled around 1878 BCE) used geometry to ensure that rents were kept fair.

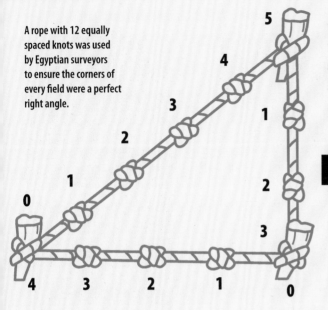

A rope with 12 equally spaced knots was used by Egyptian surveyors to ensure the corners of every field were a perfect right angle.

the shape of an old one, it was necessary to be sure that its corners really were right angles. To do this, a rope loop with 12 equidistant knots was used, as shown left. When laid out like that, the rope formed a right-angled triangle. It must be right-angled,

FRAGMENTS FROM THE PAST

We know about Egyptian mathematicians because of a few pieces of papyrus (a thick paper made from reed stems) that survived the centuries. One of these was written by a scribe named Ahmes in about 1550 BCE, and is called the Rhind Papyrus after Alexander Henry Rhind who found it in an Egyptian market in 1858. The papyrus contains only the solutions to problems (like "Find the volume of a cylindrical granary of diameter 9 and height 10"), but has no details of why the solutions give the right answers. As far as we know, the Egyptians worked out their answers through trial and error, and preserved what worked. Certainly there is no evidence that they tried to develop any mathematical proofs.

Fair rent

Plot owners were troubled by more than just the damage. Pharaoh Sesostris III had divided up the fertile land among the people in exchange for an annual rent, and he canceled rents for flood-damaged fields. But what if only part of a plot had been destroyed? Sesostris's solution was to reduce the rent in proportion to the fraction of land lost, an approach which, in part, led to the development of geometry.

Not cutting corners

To find out the rent reduction, the Egyptians first had to calculate the area of lost land. They also had to work out fractions of plot areas when land was passed on in inheritance, because fields were usually divided equally between the surviving sons. It was easy to work out the area of a rectangular plot by just multiplying the length by the width. But in order to lay out a new rectangular plot, or to check

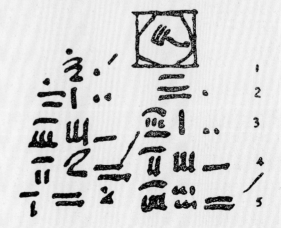

because the lengths of the sides are those given by the Pythagoras theorem, which applies only to right-angled triangles. The theorem says that for a right-angled triangle in which the shorter sides have lengths a and b, the hypotenuse (longest side) has length c, where $a^2 + b^2 = c^2$. This equation is correct for the triangle of rope, because $3^2 + 4^2 = 5^2$. The Egyptians did not know this formula, since they had no concept of using symbols nor of the ideas of "equals" or of "squared." But they did know that a triangle with sides of length 3, 4, and 5 is a right-angled one. The Egyptians also calculated the area of a circular plot, by multiplying its diameter by about 3.1. Today we know this number more accurately, and call it π ("pi").

The volume of a square-based pyramid is given by squaring the length of one of its base edges, multiplying by its height and dividing by three, or, as an equation, $V = 1/3 a^2 h$.

The volume of a truncated pyramid is $V = 1/3(a^2 + ab + b^2)h$.

How much? Ancient Egyptian architects developed geometry partly to help figure out how much stone they would need to build pyramids, great and small.

The third dimension

For building work, calculations of volume were also needed, for example, to work out the amount of stone needed to build a monument. The ancient Egyptians' now-famed fondness for pyramids made them experts on working out a building's volume and surface area from its width and height. They were even able to work out the volumes of flat-topped pyramids (described as "truncated pyramids" in mathematical terms), which are easier to build. While they used written instructions for such calculations, we can write them both more concisely and more clearly today using equations (see the diagrams on page 12).

Stuck in their ways

Although the civilization of the Egyptians flourished in many other ways for centuries after they had developed these techniques, they seem to have made no further discoveries in geometry. Having found out how to plan their buildings and measure their lands, they had no further interest in the subject.

SEE ALSO:
▸ Circles and Spheres, page 14
▸ Triangles and Trigonometry, page 56

THE PEOPLE WHO ASKED WHY

To a professional mathematician, the idea of using a formula without being able to prove it is a strange idea, like walking across a frozen lake with no idea of the thickness of the ice. But to most of us, it's perfectly normal behavior. Maybe you've used the Pythagoras theorem, or averaged six numbers by summing them and dividing by 6. These things just work, and we probably only ask "Why?" (or "Why not?") when the answer they give is clearly wrong. So it should not surprise us that neither the Egyptians nor any other civilizations before them had any notion of checking a mathematical method, other than by simply trying it out. This is true of many other things besides math, of course—not many people wonder how their cellphone works. But without a few people to do the wondering, we would still be living like the ancient Egyptians (and have no smartphones, either)! All that was to change, however, with the ancient Greeks.

The School of Athens fresco by the Renaissance artist Raphael depicts some of the great minds of ancient Greece.

Circles and Spheres

Thales of Miletus is often cited as history's first scientific thinker.

THE GREAT GREEK THINKER THALES WAS A GREAT TRAVELER TOO, leaving his native Miletus (in what is now Turkey) to explore the ancient civilization of Babylon to the east. There he was fascinated to hear that the Babylonians had discovered that a triangle inscribed in a semicircle always contains a right angle. No doubt, many other people had also been interested to hear this, but as far as we know, Thales was the first person in human history to ask seriously—of this or any other geometrical fact— "Why is this true?"

A proof from the ancient world

To prove that a triangle (T) inside a semicircle always contains a 90° angle (that is, a right angle), we first draw a line from the center of the triangle's baseline to the top of the triangle. Because this line starts from the middle of the circle, it must be a radius (r). We can label it r, and two other radii along the bottom, too.

In Thales's day—the 6th century BCE—Babylon was the most powerful state in the Middle East. The city was remembered for its immense geometric ziggurats, including the Temple of Bel, better known as the Tower of Babel mentioned in the Bible, which was rebuilt around the time of Thales.

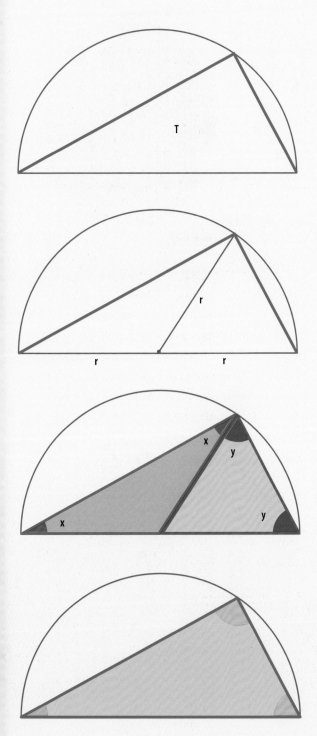

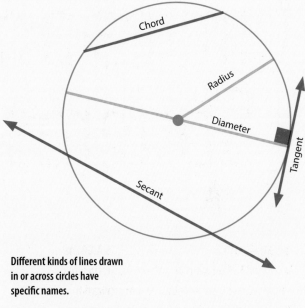

Different kinds of lines drawn in or across circles have specific names.

Now we look at the two new triangles this line has made. The red triangle has two sides of equal length (r). All such triangles (which are called isosceles triangles) have two angles that are the same. The blue triangle is also an isosceles. So both the new triangles have a pair of angles which we know are equal. We can label them as x and y. If we now forget about the new triangles and go back to the original triangle T, we can see that its three angles are x, x + y, and y. We know that any triangle's angles add up to 180°, so x + x + y + y = 180°.

We can rewrite this as: 2x + 2y = 180°
which is also: 2(x + y) = 180°
and that means: x + y = (180°)/2
which is: x + y = 90°.

That last statement completes the proof, because

the top angle of the green triangle is x + y, and now we know that this adds up to 90°.

One-sided

After the triangle, the other shape that particularly fascinates geometers, both ancient and modern, is the circle. While the triangle is a useful shape for builders and engineers, the circle was very important for all Greek thinkers because they believed it could explain the structure of the Universe. Perhaps partly because the Sun and Moon appear as circular discs, the Greeks thought that circles were natural to everything beyond the Earth, so the planets and stars must move in circles.

Circles of stone

The Greek civilization was not the only one fascinated by circles. Stonehenge, which was built over a long period starting from about 3100 BCE, is a whole series of circles which probably had both astronomical and religious meanings. The builders of Stonehenge would have found it quite simple to mark out a circle on the ground: just tying a rope to a post and walking around the post while keeping the rope

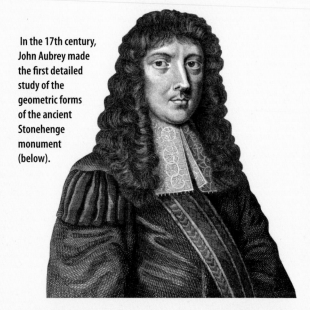

In the 17th century, John Aubrey made the first detailed study of the geometric forms of the ancient Stonehenge monument (below).

taut would do the job nicely. But laying out other details was trickier. One of Stonehenge's larger circles is marked by 56 equidistant holes, called Aubrey holes after the antiquarian John Aubrey, who discovered five of them in 1666.

Marking out

The meaning of the holes is a complete mystery, but we can guess how their positions were worked out. The planners probably started by

to see if they were the same length. A hole would have been dug at the end of each rope, establishing four equidistant primary holes. Next, a rope would have been stretched between two adjacent reference points (A and B on the diagram). Pacing the length of this rope would have allowed its midpoint (M) to be located.

Digging holes

Next, another rope would have been stretched from the center of the circle to this midpoint and extended to reach the circle (at D). Doing this for each pair of adjacent reference points would have resulted in four new holes, making a total of eight equidistant points. Trying to go beyond this using a rope is very difficult, so the next step was probably made by trial and error: between each pair of points, six new points would be estimated, giving the required total of 56 holes. The 56 Aubrey holes are quite accurately placed, with the biggest error being 19 feet between a pair rather than the ideal 16. This feat of land measurement must have been challenging and impressive at the time, but as far as we know, it did not lead to or develop from any great interest in geometry. In fact, there seems to have been little interest in geometry in Britain until the 17th century.

stretching a rope across the circle, through the central post, to give a diameter. A second rope would then have been stretched across, also through the center, and at right angles to the first. The builders might have first judged by eye if the angle looked right, and then checked by pacing along each of the four quarter circles

ROPE AROUND THE EARTH

We can't always trust our guesses when geometry is concerned, which is why it is so important to be able to prove things. For example, imagine a piece of rope that circles the Earth, just the right length so that it can lie on the surface all the way around. If you wanted to raise the height of that rope so that it hovers one yard above the surface everywhere, how much do you think you would need to add to its length?

Finding out the answer is easy. In the first case, the rope is as long as the circumference of the Earth, which is given by the equation $2\pi r$, where r is the Earth's radius, which is 6,974,880 yards. So, the length of the rope = $2 \times \pi \times 6{,}974{,}880$ yards = 43,824,464 yards. In the second case, the rope is 1 yard further away from the center of the Earth, so the radius of the rope circle is now $2 \times \pi \times (6{,}974{,}880 + 1)$ yards, which is 43,824,470 yards. So, the extra length needed is just 6 yards. Less then you thought?

Properties of a sphere

The three-dimensional version of the circle, the sphere, has also been of great interest since at least Greek times, perhaps because it is the shape of the Sun and Moon. A sphere is the smallest possible surface area for any volume, and there are many natural spheres, including planets, stars, bubbles, and eyeballs. In planets and stars, the pull of gravity from the center works to bring all the elements as close together as possible, and in any shape other than spheres, some parts would be further away than a sphere's radius. When a bubble forms, its surface is tight and resists expansion as much as possible, keeping itself as small as it can. Then, if a structure needs to be held in position by a socket and yet point in any direction, the sphere is the only possible shape, so our eyeballs are spherical. The reason is that the movable part must always fit in its holder, which has a fixed cross section, and the sphere is the only shape which has the same fixed cross section (a circle), no matter what its orientation.

Universal shape

So impressed were the ancient Greeks with the circle that they thought its shape was the only one needed to describe the Universe beyond the Moon. While they argued about whether planets (including the Earth) went around the Sun, or whether the Sun and other planets orbited the Earth, they all agreed that the paths of the planets must be circular. Medieval thinkers in Western Europe felt sure the Earth was the center of the Universe, but they agreed that circular motion was the only kind worth considering. So, from

Pluto looks circular whatever its orientation, and the only shape that can do that is a sphere.

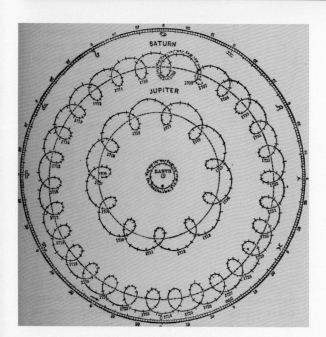

The solar system as described by Ptolemy in the 2nd century CE involved planets moving in spiral paths around Earth.

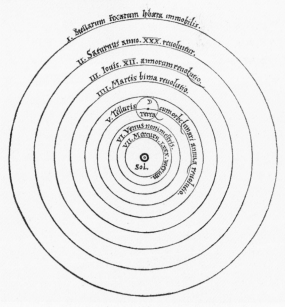

In 1543 Nicolaus Copernicus explained that the only way the observed motion of the planets made sense was if the planets, including the Earth, were moving around the Sun.

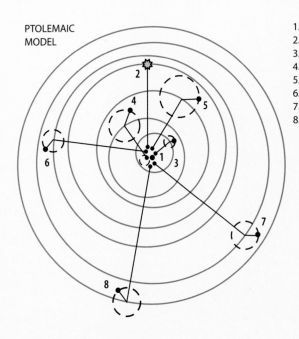

PTOLEMAIC
MODEL

1. Earth
2. Sun
3. Moon
4. Mercury
5. Venus
6. Mars
7. Jupiter
8. Saturn

COPERNICAN
MODEL

then on astronomers tried to work out, from observations of the planets in the sky, what the actual paths of the planets around the Earth could be. They soon found that simple circles could not explain the paths, so instead they tried the idea of the planets being attached to circular (or spherical) paths, which were themselves attached to other circles. Even when Copernicus showed that the Sun, not the Earth, was the center of the solar system in 1543, he still assumed that the planets must move in circles. About 70 years later, Johannes Kepler, after enormous mathematical efforts, proved that, in fact, the planets do not move in circles at all but in ellipses. If only the Greeks had been as fond of the ellipse as of the circle, astronomy would have developed much more quickly.

Copernicus disagreed with the prevailing views of his time (which followed the teaching of the Greek astronomer Ptolemy) by suggesting that the Sun, not the Earth, was at the center of the solar system. However, he agreed that the planets moved in circles. His theories on this were very complicated and involved many epicycles, or small circular orbits linked to larger ones. Above is a simplified version.

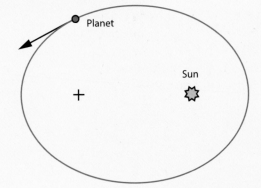

In reality, the planets' paths around the Sun are very simple indeed: the paths are ellipses.

BELTRAMI'S PSEUDOSPHERE

Because the sphere is such a simple and well-understood shape, it is ideal to use for the exploration of strange and new geometrical ideas. One of the earliest and strangest new versions of the sphere is the pseudosphere, studied by Eugenio Beltrami. Beltrami was a 19th-century Italian mathematician who combined his academic studies with political activism. He was successful in both fields, eventually becoming both the president of a venerable scientific society, Rome's Accademia dei Lincei, or the Academy of the Lynx-Eyed, and a member of the Italian senate. The pseudosphere can be thought of as the opposite of a sphere: while a sphere is a convex shape (that is, it bulges outward everywhere), a pseudosphere is concave. A sphere's surface is "closed" (you can run your fingers over it without ever encountering an edge or end or barrier), and "finite" (it has a fixed area). A pseudosphere, on the other hand, has an open surface which has an infinite area. It was once thought that our whole Universe might follow the strange laws of geometry obeyed by Beltrami's pseudosphere.

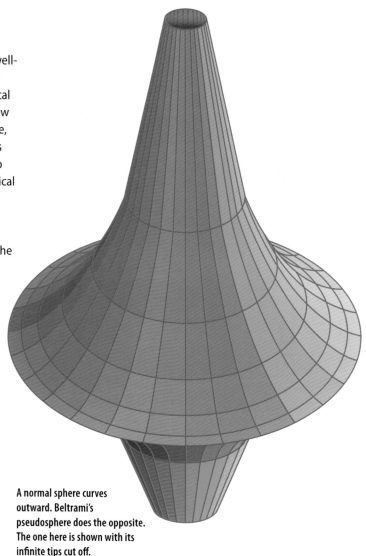

A normal sphere curves outward. Beltrami's pseudosphere does the opposite. The one here is shown with its infinite tips cut off.

SEE ALSO:
▶ Secrets of the Cone, page 38
▶ Triangles and Trigonometry, page 56

Spirals

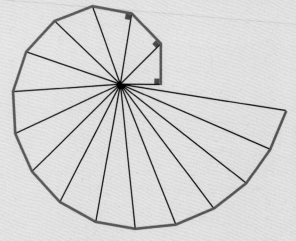

The spiral of Theodorus is made from 16 right-angled triangles that connect at a single point.

PEOPLE SEEM ALWAYS TO HAVE LOVED SPIRALS. They have been inscribed on stone structures made by various civilizations over many thousands of years. Often, spirals seem to represent endlessness, or growth, and it is easy to see why this might be. A simple, repeated motion of a stick will generate an ever-growing spiral, covering more and more space and seemingly able to go on forever.

Spirals have a long history in mathematics, too. The first person to become fascinated by them, as far as we know, was Theodorus, a Greek who lived in what is now Libya about 2,500 years ago. Although all of Theodorus's writings have been lost, Plato, another ancient Greek thinker, included him as a character in one of his dialogues (which were philosophical plays). The Theodorus character was introduced as a geometrician, astronomer, "calculator," and musician; a great man—but "not given to jest." The spiral of Theodorus, shown above, is an unusual one in that his version does not go on forever and is made of straight lines arranged into a series of right-angled triangles. Theodorus seems to have invented it to explore the square roots of numbers.

Shapes in numbers

As in many other areas of geometry, it is much easier for us to study spirals now, because we have a powerful but simple

A jug decorated with spirals from the Mycenaean civilization based on the island of Crete around 3,500 years ago.

mathematical language, or code, for them. A spiral can be thought of as the path of one object around another, in which the distance between them gradually increases. To pin down this shape, we need to describe how far around the moving object has traveled, and its distance from the stationary point in the middle. Both of these pieces of information can be measured using a choice of units. The distance might be in miles or meters, and the angular measure in degrees or radians. For spirals, radians are usually easier to use (see box, right).

Making spirals

The simplest spiral is one in which the values of the angle and the radius are the same. So, when the angle is 1 radian, the distance of the line from the center is 1 unit, when the angle is 2 radians, the distance is 2 units, and so on. Similarly, when the angle is 0.1 radians, the distance is 0.1 units and when the angle is zero, so is the distance. What emerges is a spiral like this, named after the mathematician who first studied it, Archimedes:

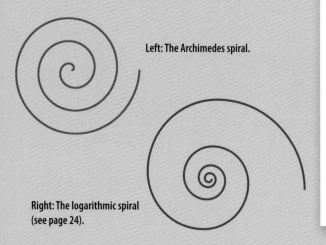

Left: The Archimedes spiral.

Right: The logarithmic spiral (see page 24).

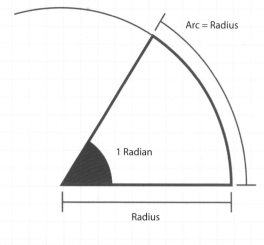

How it works

Degrees or radians?

We use 360 degrees (written as 360°) to define the angles in a circle. This is because that is what the Babylonians did thousands of years ago. They chose 360 because their numbering system was based on the number 60, and 360 is 6 x 60. (60 might seem an odd choice but it can be divided by 2, 3, 4, 5, and 6. You cannot do that with 10, which is only divisible by 2 and 5). However, we could just as well decide that there are 100, 120, or 1024 angle units in a circle. A more natural way is to choose the most basic element of a circle, its radius. If we mark off a distance along the edge of a circle which is equal in length to the radius of that circle, the resulting shape, like a large pizza slice, would have the same angle whatever size of circle we chose. So we can use this angle as our unit and call it the radian. Since there are 2π radii in a circle, it follows that there must be 2π radians in it, too. So 2π radians = one full circle = 360° which means that 1 radian = (360°)/2π = about 57.3°. Often, angles in radians are written in terms of π, so a right angle may be labeled as π/2.

Arc = Radius

1 Radian

Radius

Miraculous spirals

Some spirals are found in the natural world, and one of these so impressed the great 17th-century Swiss mathematician Jacob Bernoulli that he called it the miraculous spiral. It can be found in many galaxies, snail shells, rams' horns, elephants' tusks, and in the patterns of seeds on the head of every sunflower. Peregrine falcons even fly in miraculous spirals!

The miraculous spiral is better known today as the logarithmic spiral (see diagram, page 23). While the curves of the Archimedean spiral keep the same distance apart as you go outward, in a logarithmic spiral the distance between the curves

steadily increases. What fascinated Bernoulli about this spiral is that it is "self-similar," which means that however much we enlarge or reduce it, its shape remains the same. More than a century after Bernoulli's death, the subject of such shapes, called fractals, became a major area of mathematical research. Bernoulli liked the logarithmic spiral so much that he asked for one to be inscribed on his tomb. Sadly, though, the stonemason was no expert on spirals and carved an Archimedean one instead!

Natural shape

But why should this spiral be so common in nature? Another ancient Greek, Aristotle, gave part of the answer in about 350 BCE, when he noticed that some plants and animals grow by adding units that are always the same shape, but a larger size. Such shapes are called gnomons, and if we use a triangle as a gnomon we can build a shape which looks at least a bit like a spiral. A sea creature called a nautilus has a shell in the shape of a logarithmic spiral because it builds it by adding

Left: The inscription on Jacob Bernoulli's tomb includes a spiral at the base —just not the one that the great mathematician wanted.

Right: A rough spiral can be made by adding triangles which are each twice the area of the previous one.

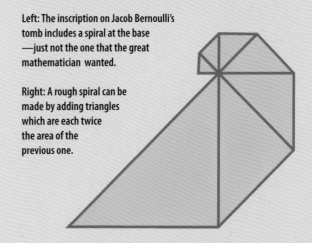

Flying in logarithmic spirals solves a problem for peregrine falcons. Because their eyes are in the sides of their heads, their forward vision is poor. The birds see best at an angle of about 40° from straight ahead. So, when hunting for prey in front of it, a falcon should ideally hold its head at an angle of 40°. But this would increase wind resistance. Flying in a logarithmic spiral means the falcon can keep its head straight while maintaining the best viewing angle all the time.

A nautilus is a relative of the squid. It uses the smaller, gas-filled chambers in its spiral shell to help it float.

The peregrine falcon flies in a logarithmic spiral to ensure it can look for prey in all directions.

"chambers," each the same shape as the previous one, but larger. The nautilus lives in each chamber in turn, adding a new one to move into once it has grown too large to fit comfortably in the old one.

Constant angle

Self-similarity explains most of the natural occurrences of the spiral, and another feature of the spiral may explain the rest. The logarithmic spiral is "equiangular," which means that, if you draw any line through the spiral from the center, it will pass through the spiral line at the same angle.

SEE ALSO:
▶ Archimedes Applies Geometry, page 52
▶ Fractals, page 166

The Mathematics of Beauty

CAN BEAUTY BE CAPTURED MATHEMATICALLY? Some say it can, by a single shape: a rectangle in which one side is about 61.8 percent longer than the other. (If one side is 10 inches long, the other will be 16.18 inches.) This 1 to (about) 1.618 relationship is called the golden ratio, and was first made famous by Phidias, the greatest sculptor of ancient Greece. Phidias is thought to have helped plan the Parthenon, a famous temple built on the great hill of Athens, in which the golden ratio appears many times.

After this, the golden ratio became very popular in art and architecture, and today it is the shape of credit cards. Phidias was not so popular. Becoming involved in political intrigue, he was accused of

The golden ratio is present in many parts of the Parthenon.

How it works

The golden spiral

A rectangle with sides in the golden ratio is known as a golden rectangle. To find out whether a rectangle is golden or not, there is no need to measure its sides. if a line is drawn across the rectangle to form a square, the remaining rectangle should be the same shape as the original—another golden rectangle. Repeating this process by drawing a square in the new rectangle again produces a golden rectangle, and this can be continued indefinitely. The shape that forms has a number of properties; it is a fractal (see page 166) and it can be used to draw a kind of spiral, called the golden spiral.

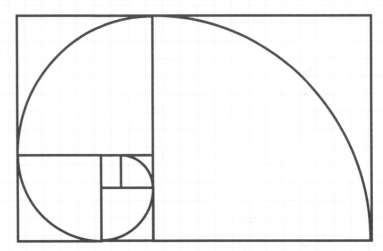

stealing gold intended for the making of a statue, and he was exiled and executed soon after. His initial, which in Greek is written φ and pronounced "fie", is now used to symbolize the golden ratio.

In words

The first person to explore the properties of the golden ratio was the ancient Greek Euclid, perhaps the greatest geometer of all. He gave it a precise definition:

TAKE A LINE AND DIVIDE IT INTO TWO LENGTHS SO THAT THE RATIO OF THE WHOLE LINE TO THE LONGER ONE IS THE RATIO OF THE LONGER TO THE SHORTER.

This definition has no numbers in it, because the Greek mathematicians had no interest in including numbers with their geometrical figures. The ratio appears in geometrical shapes as well as in art. In particular, the five-pointed star, or pentagram, contains many examples of the ratio:

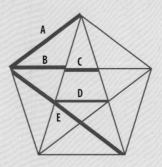

$$A \div B = B \div C = D \div C = (B+C) \div B = E \div A = \varphi$$

In numbers

To find out more about φ, we must go further than Euclid, and explore its numerical value. The equation for φ is a very strange one in that it defines φ in terms of itself. This kind of thing is not usually a good idea. After all, defining a rectangle as an oblong, or an oval as an ellipse, is of no use at all. But for φ, it works:

$$\varphi = 1 + 1/\varphi$$

In numbers, this becomes $\varphi = 1 + 1/(1.61803\ldots)$ We can write this in another way too, as an endless (or "continued") fraction:

$$\varphi = 1 + \cfrac{1}{1 + \cfrac{1}{1 + \cfrac{1}{1 + \cfrac{1}{1 + \ldots}}}}$$

The fact that this goes on forever means that φ cannot exactly be expressed as a ratio of numbers, so it is known as an irrational number (see page 46).

In nature

This property also explains why φ turns up in naturally beautiful objects, such as sunflowers. When a sunflower grows, it gradually adds new seeds at the edge of the flower. This is similar to laying out counters in a spiral sequence, starting from the center. One way to do this would be to make a quarter-turn between counters, which would mean placing them at 0.0, 0.25, 0.5, 0.75 fractions of a circle, like diagram A, below.

Using the golden ratio

However, the sunflower needs to be fully covered with seeds, so this is not much good. Making smaller turns helps a little. Diagram B is what is created with turns that are each one-tenth of a circle (so, the seeds are placed at 0.1, 0.2, 0.3, and so on of a full circle). This is still not ideal. In fact, whether we try thirds, fifths, hundredths, or even thousandths of a circle, we still get gaps. It turns out that the way to cover the seed head is not to use any of these so-called rational fractions, but to use φ. Then we get diagram C:

A   B   C

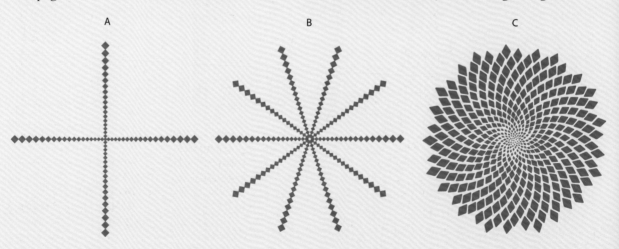

Link with Fibonacci

Hidden in this pattern is a sequence of numbers called the Fibonacci sequence. This sequence is made by writing two 1s and then writing their sum, 2, then adding this 2 to the previous number, to get 3, then adding this to the previous number, to get 5, and so on. That process makes this sequence:

1, 1, 2, 3, 5, 8, 13, 21…

If we try dividing these numbers in pairs, we find answers that gradually get closer and closer together, as shown in the third column here. The answers appear to be homing in on and getting ever closer to something familiar.

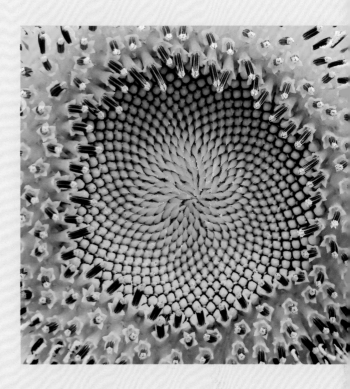

A	B	B/A
2	3	1.5
3	5	1.66666666…
5	8	1.6
8	13	1.625
13	21	1.615384615…
and so on along the sequence, then		
144	233	1.618055556…
233	377	1.618025751…
and so on along the sequence		

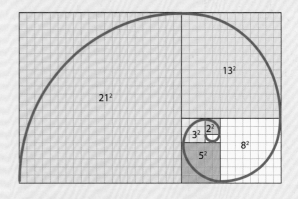

The square elements of a series of divided golden rectangles—making a golden spiral—have Fibonacci number dimensions.

We can see these Fibonacci numbers, as they are called, in real sunflowers, and we can also see them in the golden spiral.

SEE ALSO:
▸ Spirals, page 22
▸ The Geometry of Architecture, page 106

The Shapes of Perfection

LIKE MANY OTHER CIVILIZATIONS, THE GREEKS WERE FASCINATED BY THE IDEA OF PERFECTION. However, unlike other ancients (but like scientists today), they were also interested in finding out what everything is made of.

It seemed obvious to the Greeks that perfection is very rare in the world around us, if it exists at all. But geometry did offer some possibilities in the forms of polyhedrons, solid shapes with flat sides. The Greeks were especially interested in five of them: the cube, tetrahedron, octahedron, icosahedron, and dodecahedron. These have sides that are all identical polygons (a flat shape with straight edges) which all meet at the same

angles, so these solid shapes were called "perfect" polyhedra. The perfect solids were probably grouped together in about 420 BCE by Theaetetus, who was a student of Theodorus (see page 22) and who also proved that there can only be five of them. Rather unfairly, they are called the Platonic solids after the philosopher Plato, who was very fond of them. Plato is also the source of all that we know about Theaetetus, which is not much: Plato said he had bulging eyes, a snub nose, and died of his wounds after a battle.

A more perfect world

The shapes are named after Plato because of their importance in his theories. In his view, the things we see around us are no more than imperfect copies of perfect things called "forms," which cannot be sensed, but can be understood by reason. So, any ball, orange, or other spherical object is a rough copy of an actual, invisible,

The five regular solids are named after the philosopher Plato, who believed they explained the properties of all matter.

How it works

Everything is a triangle
All convex polygons (where internal angles are always less than 180°) can be divided up into triangles. To see this, just draw any polygon, pick a corner, and draw straight lines from it to all other corners. Platonic solids are made of polygons, so Plato thought of them as being made of triangles as well.

perfect sphere, and the pyramids of Egypt are rough copies of a perfect version. Up to a point, this idea makes good sense. When we sketch a circle there is no doubt that it is an imperfect one, and other people will recognize both that it is not a perfect circle and that it is intended to represent one. We could make a more perfect circle by using a compass, but again we would not expect this one to be quite correct either.

Building blocks

For Plato, the fact that we know what a perfect circle (for instance) would be like, even though no actual circles are perfect, meant that we must have some inborn sense of a geometric world: a perfect world of forms. Like so many ancient Greek thinkers, Plato scorned experiments and measurements, and preferred to ponder and

discuss his ideas, and he became so keen on his idea of the mysterious forms that in the end he decided that everything in the world is a copy of a perfect, invisible form, from circles and cubes to trees and hats. But, for Plato the five Platonic solids were more important than spheres or pyramids because they can be constructed from triangles, which he regarded as the building-blocks of nature, much as we regard atoms today.

Elemental forms

Plato also thought that the structure of matter could be explained by the Platonic solids. By his time, many thinkers believed that everything was made of four elements: earth, air, fire, and water. (Thales seems to have come up with the first version of this theory, in which he suggested that everything was made of water). Some added a fifth

HOW MANY PLATONIC SOLIDS CAN THERE BE?

Platonic solids are regular polyhedra, which means that the angles of the shape are always equal, as are its lengths and areas. A flat side is called a face, and where two sides meet is known as an edge. A corner where more than two edges meet is known as a vertex (the plural is vertices). To meet the criteria:

1. In any solid shape with flat faces, at least three of those faces must meet at each vertex. Try to think of a solid where fewer than three flat faces meet. There are none.

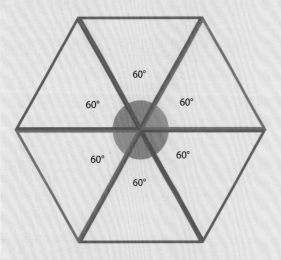

2. Every vertex has a total internal angle, which is the sum of the angles between the edges that meet there. The maximum total angle of a vertex is 360° (a full circle), but that is only possible if the faces are all on the same plane—in other words if the shape they make is flat. The diagram above shows how six equilateral triangles, triangles with threes sides of the same length, meet as a flat vertex with a total

angle of 6 x 60° = 360°. However, this maximum can never apply once a vertex moves into the third dimension.

3. However, for a solid shape, we do need to imagine triangles that meet at a raised point that exists in three dimensions. The total internal angle of the point must be less than 360°, and as the point is raised higher, that total angle gets smaller. Also, in a Platonic solid, by definition, the shapes that meet are identical, so each of their individual angles must be less than 120°. This rules out a number of shapes as being allowed to form Platonic solids. A hexagon, for instance, has vertices which have angles of 120°, so it must be ruled out, and so must all polygons with more than six sides. So, a Platonic solid must be based on shapes with fewer than six sides, which means triangles, squares, and pentagons.

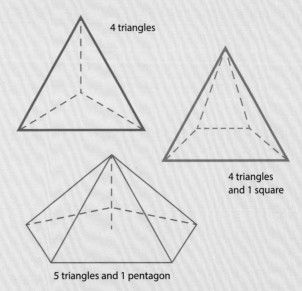

4 triangles

4 triangles and 1 square

5 triangles and 1 pentagon

4. Each vertex of an equilateral triangle is 60°, so only 3, 4, or 5 such triangles can meet to form a total angle less than 360° (as shown above). These three possibilities give the tetrahedron, octahedron, and icosahedron.

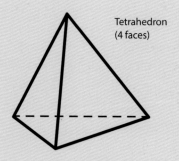

Tetrahedron
(4 faces)

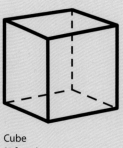

Cube
(6 faces)

5. That exhausts the possibilities of triangles, so we move on to consider squares. Each vertex of a square is 90°, so only 3 squares can meet to form a total angle less than 360°, and this forms a cube when three more squares are added.

Octahedron
(8 faces)

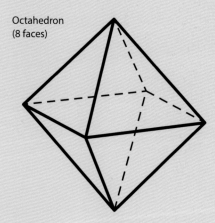

6. The cube is the only possible shape made from squares, so we turn to the last possible polygon, the pentagon. Each vertex of a pentagon is 108° so again only three pentagons can meet, and the resulting shape is a dodecahedron, once nine more pentagons have been added.

Dodecahedron
(12 faces)

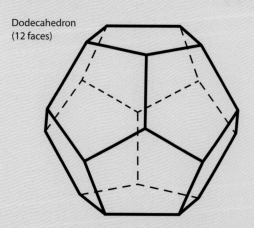

Icosahedron
(20 faces)

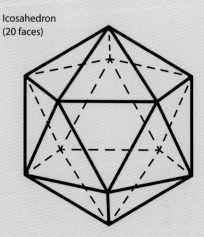

7. There are no further regular polygons we can use, so these five shapes are the only Platonic solids that can exist.

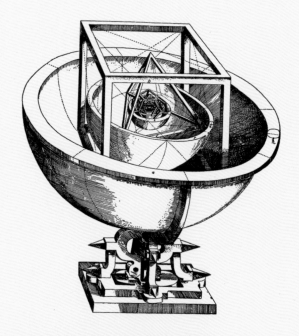

Johannes Kepler, who later figured out how planets move in their orbits, attempted to construct a model of the solar system using the five Platonic solids surrounded by an outer sphere.

to develop the idea of atoms, the enormous popularity of the Platonic solids long after the ancient Greek era was to cause problems. Johannes Kepler, who was one of the greatest geometers and astronomers of the 16th century, believed that he could explain the distances of the planets from the Sun by arranging all five of the Platonic solids inside one another—one for each planet (only five planets were known in those days). It is true that he got approximately the right answers by doing this, but there are so many ways to sequence the shapes that he was bound to get roughly what

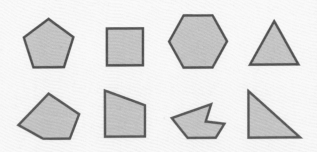

The blue polygons are regular, with edges of an equal length meeting at points of equal angles. The red shapes are irregular.

element, the quintessence or aether, which exists only beyond the orbit of the Moon and of which the Sun and stars were made. The fact that there are five Platonic solids and five elements encouraged Plato to pair them off, and he and other Greeks attempted to explain the qualities of each element by the shape of the elements which made it up. So, they thought Earth was made of cubes, which stack together nicely, while fire was made of tetrahedral atoms, and it burned when touched because the tetrahedra are so pointy and sharp, and so on.

Dangerous Legacy

While Theaetetus's proof was a mathematical triumph, and Plato's theories may have helped

The yellow shapes are convex; their internal angles are all 180° or less. The green shapes are concave; they have internal angles over 180°.

he wanted. And almost any other shapes would have worked just as well!

Polygons and polyhedra

The polygon sides of Platonic solids all have sides of the same length, so are called regular polygons. Irregular polygons have sides of different lengths. So a right-angled triangle is an irregular polygon, while an equilateral triangle, which has sides all the same length, is a regular polygon. Another distinction is between convex and concave polygons. The difference is that the internal angles of a convex polygon are all 180° or less. Any line drawn from one corner to another in a convex

does not leave the inside of the shape. By contrast, at least one of a concave polygon's internal angles is more than 180°. If we use more than one kind of polygon to construct a solid, but still require that all its angles are the same, we get the 13 Archimedean solids (see overleaf). However, if a solid is neither Archimedean nor Platonic, there are three other possibilities. It could be a prism, which is a pair of polygons connected by rectangles, or an antiprism (a pair of polygons connected by triangles). Finally, it could be a Johnson solid, named for mathematician Norman Johnson. These are made of more than one kind of polygon, and their internal angles vary. A square-based pyramid is a Johnson solid, and there are 91 others described so far!

SHAPE FAMILIES

By changing the definition of a polygon, different families of shapes emerge. For example, a square is defined as having four equal sides and four right angles. If we remove the rule that the sides must be equal, the result is a rectangle. If we also remove the rule about the angles being the same, but add a new rule that opposite sides are parallel, we can get a parallelogram. But a rectangle does not have to have unequal sides, so a square is a kind of rectangle.

Similarly, a parallelogram does not have to have unequal angles, so a rectangle is also a kind of parallelogram. The diagram to the right classifies different kinds of quadrilaterals, or shapes with four sides. A small square at a corner means the shape contains a right angle. If the same mark appears on two (or four) sides of a shape, then that shape must have those two (or four) sides equal.

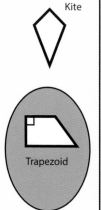

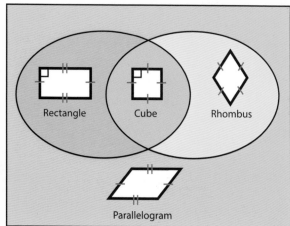

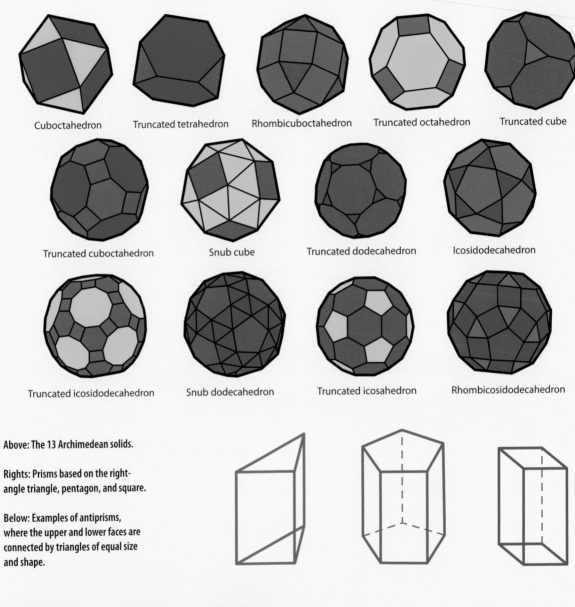

Cuboctahedron   Truncated tetrahedron   Rhombicuboctahedron   Truncated octahedron   Truncated cube

Truncated cuboctahedron   Snub cube   Truncated dodecahedron   Icosidodecahedron

Truncated icosidodecahedron   Snub dodecahedron   Truncated icosahedron   Rhombicosidodecahedron

Above: The 13 Archimedean solids.

Rights: Prisms based on the right-angle triangle, pentagon, and square.

Below: Examples of antiprisms, where the upper and lower faces are connected by triangles of equal size and shape.

SEE ALSO:
▸ Filling Space, page 92
▸ Crystals, page 136

THE MYSTERY OF MELANCHOLIA

Many artists have used polyhedra in their work, but they seem to have especially fascinated the German Albrecht Dürer, who not only wrote about them but invented a new one: the snub cube. Shapes like this are hard to draw clearly, and it was Dürer who was first to draw "nets" of polyhedra. A net is a two-dimensional shape that can be cut out, and then folded and stuck together to produce a three-dimensional solid. So, while we cannot be quite sure about the shape of the snub cube from the picture below (maybe there is an unseen hexagon at the back), we know what Dürer had in mind because he provided its net. However, in 1514 he made a strange etching called *Melancholia* which includes a polyhedron drawn in such a way that it cannot be definitely identified. People have argued for centuries about what it is and what it means, for, perhaps to maintain the mystery, Dürer provided no net for the shape.

The snub cube as described by Dürer is one of the Archimedean solids.

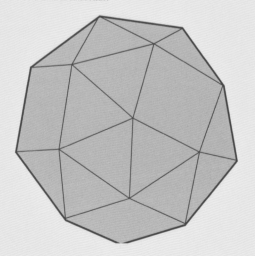

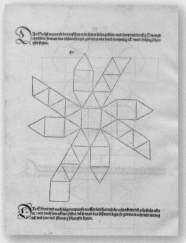

Above: As well as the mystery polygon, Albrecht Dürer's *Melancholia* contains other mathematical features, including a magic square.

Left: Dürer's original net of the snub cube.

Secrets of the Cone

IF A PAIR OF CONES ARE ARRANGED POINT TO POINT, AND THEN CUT THROUGH WITH A STRAIGHT LINE, then the edges of the cut faces will have one of three shapes, depending on the angle of the cut. A horizontal cut will form a circle, while steeper cuts will form ellipses, parabolas, or hyperbolas. Hyperbolas appear in pairs because the cut will pass through both cones.

Cones were common shapes in ancient arts and crafts, such as this ancient Egyptian vase (above) and this religious offering from Sumeria (right) with sacred script written on a clay cone.

These shapes are called the conic sections, and they interested early mathematicians almost as much as the Platonic solids. Then, as other areas of mathematics developed, such as algebra, the significance of the conic sections also grew.

On cones

The conic sections were first studied in about 350 BCE by Menaechmus, a mathematician about

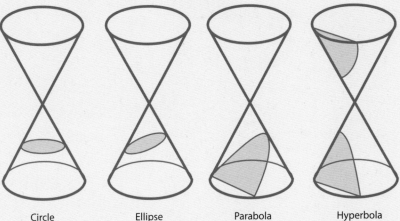

Circle   Ellipse   Parabola   Hyperbola

NEW DEVICES

Drawing conic sections was not easy until graphics software was developed in the middle of the 20th century. So, in the 1650s, a Dutch mathematician called Frans van Schooten set to work to make life easier for his colleagues. He invented the method of drawing an ellipse using two pins and string, but he could not find such a simple way to construct parabolas (below, left) or hyperbolas (right). Instead, he managed to construct instruments from hinged rods to do the job—and he invented one for ellipses, too (center).

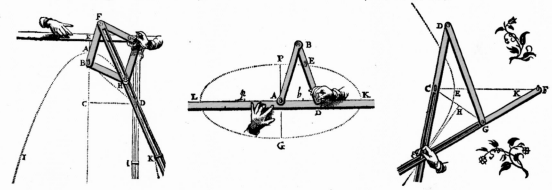

whom we know almost nothing for sure, though according to some early writers he was a tutor of Alexander the Great. Supposedly, Alexander asked Menaechmus for a swift and simple way to master geometry, to which Menaechmus replied "O King, for traveling over the country there are royal roads and roads for common citizens, but in geometry there is one road for all!" Sadly, we don't know Alexander's reply.

Apollonius

Because all of his writings are lost, we know little about what Menaechmus discovered about the conic shapes. The man who literally wrote the book on the subject, an eight-volume master-piece, was Apollonius of Perga, who lived about a century after Menaechmus. Most unusually for an ancient Greek individual, we know a fair amount about Apollonius, because he wrote a short auto-biography. In this he recorded that he wrote the book at the request of a fellow geometer called Naucrates, who was staying with Apollonius for a while. Apollonius just managed to finish the whole eight volumes before it was time for Naucrates to leave—presumably Naucrates visited for quite a while. As a result, Apollonius worried that he had no time to check the details.

Ellipses

All planets and most comets orbit the Sun in ellipses, and all moons orbit planets in ellipses, too. (This was discovered in the 17th century by Johannes Kepler.) These ellipses range from being very nearly circular to being very flattened

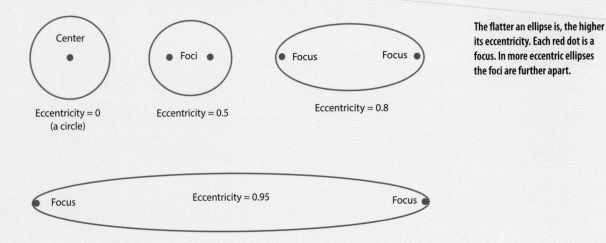

Center

Eccentricity = 0
(a circle)

Foci

Eccentricity = 0.5

Focus   Focus

Eccentricity = 0.8

The flatter an ellipse is, the higher its eccentricity. Each red dot is a focus. In more eccentric ellipses the foci are further apart.

Focus   Eccentricity = 0.95   Focus

indeed. The shape of an ellipse is measured by its eccentricity (see above). An ellipse can be made by tying string between two pins in a board, pulling the string taut with a pencil, then drawing a loop. The positions of the pins are the foci (the singular is focus) of the ellipse. When a satellite orbits the Earth, it always does so in an ellipse, with the Earth at one focus. The same is true of any other small object orbiting a much larger one. All conic sections have at least one focus point. In the circle, for example, the focus is the center point.

Vertex and directrix

While circles and ellipses are closed shapes that surround a finite area, and are each composed of a line of a particular length, parabolas and

hyperbolas are endless lines dividing infinite space. Parabolas divide space into two, and hyperbolas divide space into three areas. Pinning these curves down so we can understand them is more difficult. Both of them have foci, but we need something else to measure against. This is a line called the directrix, which is a set distance from the sharpest curve of the parabola or hyperbola. The point on the curve closest to the directrix is called the vertex.

The three parameters or measuring points of a parabola are the focus, vertex, and directrix.

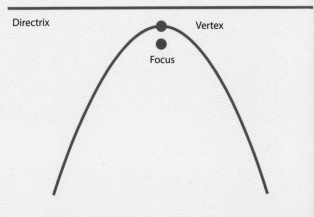

Directrix   Vertex

Focus

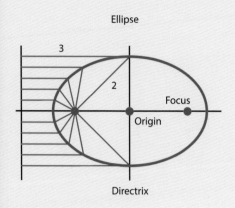

Ellipse

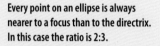

Every point on an ellipse is always nearer to a focus than to the directrix. In this case the ratio is 2:3.

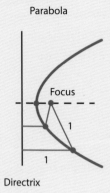

Parabola

The distance to the focus and directrix is the same from every point on a parabola.

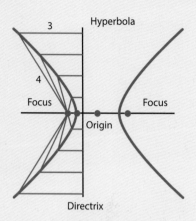

Hyperbola

In a hyperbola every point is nearer to the directrix than to the focus. In this case the ratio is 4:3.

No cone needed

While the conic sections are interesting and useful, the idea of making them by cutting a double cone may seem rather weird as well as inconvenient in practice. Fortunately, the ideas of directrixes and foci offer a simpler way to produce the sections. If a line is drawn so that the distance to the focus is half of that to the directrix the shape that will be drawn is an ellipse. If the pen is always the same distance from the focus and the directrix, the result is a parabola, and if the distance to the focus is twice that to the directrix, a hyperbola results. The process of plotting and drawing these curves is not easy but certainly something that can be done by hand.

Trajectory through space

Parabolas are of as much interest to space scientists as ellipses. If a smaller object sweeps close by another larger one without going into orbit—in an ellipse—the gravity of the second

object will make the first move in a parabolic path. Also, telescope mirrors are parabolas when seen in profile, as is easy to see on a satellite dish or radio telescope. The reason is that a parabolic dish will reflect all the light (or other kinds of waves) from a distant object to its focus. (Another way of saying this is that the dish focuses the light.) To capture the image being produced, a device that can detect light—a camera or an eye, for example—is placed at the focus.

Ballistic path

A bullet or artillery shell moves in an approximately parabolic path through the air. The parabola is approximate because the resistance of the air holds the projectile back, narrowing the parabola a little. On the airless Moon, projectiles travel in truly parabolic trajectories, unless they are fast enough to enter orbit, in which case they will take up elliptical or circular paths. This was worked out by Isaac

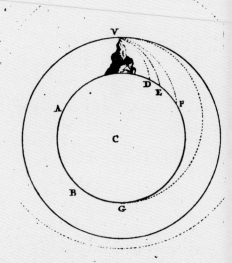

Fig. 1

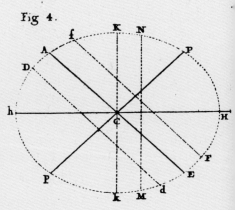

Fig 4.

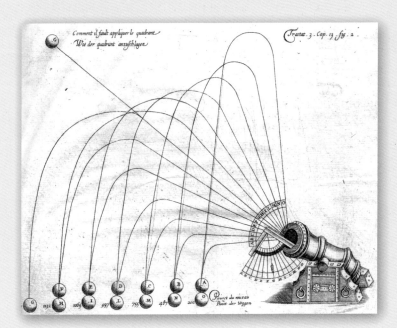

The old idea that a cannonball has a straight-line flight path (top) was replaced with the correct, curved path (bottom). Galileo's discovery of this was used to develop techniques for targeting artillery.

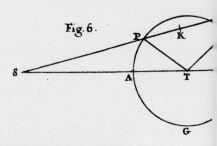

Fig. 6.

Newton filled the back pages of *Principia*, his 1687 book on gravity and motion, with illustrations, including Newton's Cannon (top) which showed elliptical and parabolic paths of flying objects.

OUR ELLIPSOIDAL HOME

Just as a three-dimensional version of a circle is a sphere, so the ellipse has a three-dimensional version, the ellipsoid or spheroid. Again, just as there are many differently shaped ellipses, so there are also many different spheroids. The two most important are the prolate and oblate types.

An oblate spheroid is formed by rotating an ellipse around its short axis, and is the shape of an orange—and also of the Earth. The Earth's gravity tends to pull it into a spherical shape, but its spin modifies this shape by making it bulge at the equator. This effect is the same one which causes a weight on a string that is whirled around to rise in the air, and will also tend to pull your arms outward if you spin on the spot—perhaps at an ice rink. Jupiter is the most oblate planet we know: it is about 7 percent wider than the distance between its poles. If an ellipse is rotated around its long axis, a prolate spheroid is formed. Footballs and some eggs and fruits have this shape.

Perfect spheres are rare in nature. Instead, most spheres are really ellipsoids, either prolate (top) or oblate like Earth, which bulges in the middle and is flattened at the poles.

Newton. He illustrated the idea in his most famous book, *Principia*, as a cannon ball fired at different speeds from a cannon.

Artillery targeting

Some decades before Newton's work, in the 1590s, Italian scientist Galileo Galilei had already worked out that projectiles on Earth fly in curves which would be parabolas if there were no air resistance. It was very useful to Galileo that the geometry of the parabola was known, because the mathematical tools to handle moving objects had not been invented (it was Newton who developed them). Galileo's discovery broke with the views of the ancient Greeks, which most scientists still accepted even then. Their version was that a projectile traveled in a straight line until the force that sent it on its way was used up, and then it fell straight down. The truth of the matter was important because it helped to target cannons to hit a distant target. Hyperbolas are not common in nature, but some high-speed comets travel in hyperbolic paths around the Sun.

SEE ALSO:
▶ Circles and Spheres, page 14
▶ Geometry + Algebra, page 98

Euclid's Revolution

Euclid of Alexandria is thought to have lived in the 4th and 3rd centuries BCE.

NO ONE IS MORE IMPORTANT TO THE HISTORY OF GEOMETRY THAN EUCLID. The subject as we understand it today might not even exist without him. As with so many ancient Greeks, we know hardly a thing about him, but copies of *Elements*, his greatest work, have survived. *Elements* is one of the most important books ever written. For nearly 2,000 years, schoolchildren and scholars alike were taught mathematics using Euclid's *Elements*.

Elements is important for two reasons. Firstly it collected almost all of the geometrical knowledge available in Euclid's time, including 465 theorems and constructions, and it also helped define the way mathematics should be done. Rather than simply stating theorems, Euclid proved them— all of them. By a proof, Euclid meant something that is quite certain. For him, it was not enough simply to measure the angles inside a selection of triangles and conclude that, since they always added up to two right angles, the same must be true of every possible triangle. For one thing, how could anyone know that tomorrow a cunning mathematician would not find a triangle whose angles added up to three right angles?

Something like this actually happened in the 18th century (see more on page 130)!

Mathematical foundations

Secondly, even with modern instruments, no one can draw a perfect triangle. In Euclid's time people usually drew their diagrams on sand, so most triangles were very far from perfect, and angles could not be measured with great accuracy, either. Key to Euclid's genius is the idea of an axiom (which he called a postulate), a fundamental statement on which many theorems depend. Euclid only needed five axioms, and four seem so obvious that they may seem hardly worth stating. They say: 1) Given any two points, you can draw a straight line between them.

BOOK I. PROP. XV. THEOR. 15

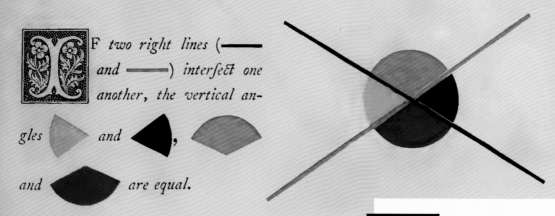

*T*F *two right lines* (——— *and* ———) *interſect one another, the vertical an-gles* and , , and are equal.

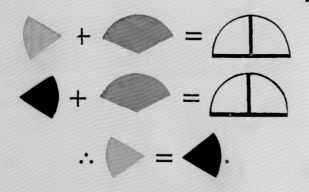

THEOREMS

A theorem is a statement that can be proved. Before the Greeks, no one seems to have even thought of proving mathematical statements, and, before Euclid, not all Greeks thought that mattered, either. Euclid's determination to prove every theorem was new, but it is fundamental to all areas of mathematics and science today.

In 1847, a version of *Elements* using colored shapes was published. This page illustrates one of Euclid's simpler proofs.

In the ſame manner it may be ſhown that

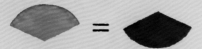

Q. E. D.

CONSTRUCTIONS

For Euclid, a construction means a way to do something using just two simple tools, a compass (a pair of compasses) and a straight edge (or ruler). He limited himself just to these tools because they were the only ones used by all Greek mathematicians and because it was easy to check that they were properly made. The straight edge would not have been marked with numbers, like a modern ruler, because mathematicians in those days did not want to rely on numbers. Centuries before, Pythagoras (or one of his followers) had found numbers that can neither be written down, nor marked on a ruler, nor expressed as a fraction. One of these "irrational" numbers is the square root of 2. It is roughly equal to 1.41421356, but not exactly. No one knows, nor can ever find out, exactly what its value is. It was to be many centuries before mathematicians learned how to deal with irrational numbers.

In this construction, Euclid shows how to halve an angle.

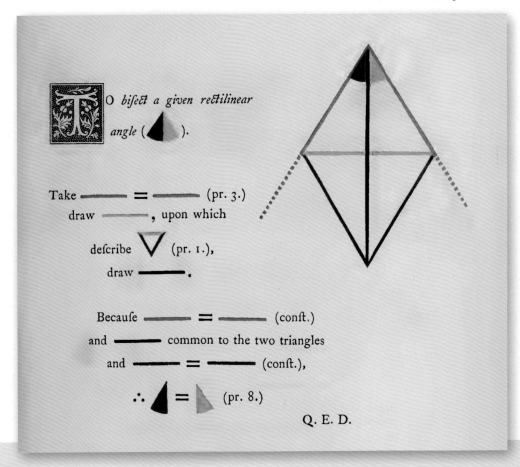

This fragment of papyrus written in ancient Greek is the oldest existing copy of Euclid's *Elements*, dating from 100 CE.

2) Any line may be extended as far as you like. 3) A circle may be drawn with a center wherever you want, and with a radius of whatever length you choose. 4) All right angles are the same.

Rules of the game

Why did Euclid bother? One reason was to avoid arguments that start "I just don't believe it." Take Eratosthenes, for example, who used geometrical figures that extended all the way to the Sun (see page 50). If someone were to say "I don't believe geometry can be used like that. No line could ever really reach the Sun, so I don't agree with your conclusions," then Euclid could respond "If you accept my axioms, including the one that states that any line may be extended as far as

These pages show the full proposition in the fragment shown above.

you like, then you must accept my conclusions. Whether you accept the axioms or not is up to you, I do not attempt to prove them. In either case, there is nothing for us to be arguing about." So in a way the axioms are like the rules of the geometry game—if you want to play Euclid's game, you will learn a great deal, but only if you accept the rules. Another important point

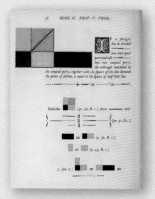

about the axioms is that, together with some clear definitions, they are the only things Euclid needed. Even the most complex proofs were based on them, by taking many small, logical steps, each of which can be carefully checked. This means that those complicated proofs are just as certain as the axioms are.

Not perfect

There is no doubt that the world is a better place thanks to Euclid, but there are three problems with the enormous influence of the book. The first, and least important, problem is that there are a few errors in *Elements*. But this is not entirely a bad thing. If even the great Euclid can go wrong, it shows how careful others must be. And the fact that the errors can be proved wrong by using Euclid's own axioms and methods demonstrates just how powerful those methods are. Finally, it is a very useful lesson, too, in not believing what anyone says, just on the basis of their fame or brilliance: only proof will do. The second problem is that some of Euclid's proofs

are far more complex than necessary, while others are not explained as clearly as they should be. This can make the book hard to use as a teaching aid, and is the main reason why people nowadays do not learn geometry from it. (By far the easiest way to learn geometry today is through a video-based system like YouTube, since it is natural to follow a proof through changing diagrams.)

A big problem

The third problem with *Elements* is the most significant. Euclid's fifth axiom is much less obvious than the rest. There are several ways to explain it, but one of the simplest is "Given a straight line, there is only one other kind of straight line which is parallel to it." This seems fairly clear and reasonable. Draw a line with a ruler and draw another one as parallel as you can to it. You can draw lots of other lines parallel to it, but they will all be parallel to each other, too— that is, they are all the same "kind of straight line." It's hard to imagine drawing a line which is not parallel to any of your drawn lines, but

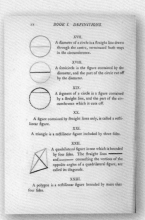
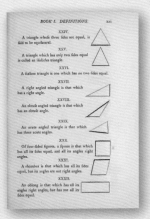

which is parallel to the original one. Even so, you could say that this is like the statement that all triangles must have angles that add up to two right angles; just because no one has found a triangle whose angles add up to more or less than two right angles does not prove that no such triangle exists. This triangle statement is a theorem that needs to be proved, not an axiom. It seems to have struck Euclid himself that his fifth axiom was not quite as satisfactory as the rest, and he and many others attempted to show that it could actually be proved on the basis of the other four. But no one managed it. There is a very good reason for this, which is explained on page 68.

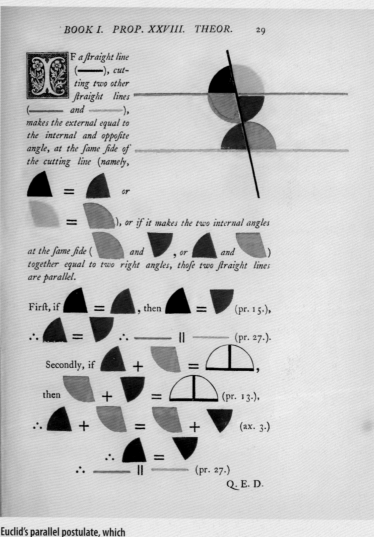

Euclid's parallel postulate, which he could not prove, led to a revolution in geometry 2,200 years after it was written.

The early pages of Euclid's *Elements* contain dozens of definitions based on the five axioms.

SEE ALSO:
▶ Triangles and Trigonometry, page 56
▶ Non-Euclidean Geometry, page 130

Eratosthenes Measures the Earth

ERATOSTHENES WAS A MAN AHEAD OF HIS TIME. He was determined to study the entire Earth and made one of the earliest maps of it; he is remembered as the founder of geography. He was also unusual for his time in wanting to pin down facts by using numbers. So, while many people before him had written about the past, he was the first to try to develop histories with accurate dates. And he was determined to put a number to the size of the Earth, too.

Since the Earth was trillions of times bigger than Eratosthenes, any kind of direct measurement was unlikely to work, especially since it would involve traveling around the world at a time (about 250 BCE) when no one knew what most of it was like or even whether Earth was definitely round. Some people thought a traveler would fall off the edge of the world. (In 1799 an expedition covered a quarter of Earth's circumference.)

Using geometry

Eratosthenes' method is very simple but still very clever. For one thing, it assumes that the Sun is so far away that it might as well be at infinity. Secondly, it requires the simultaneous measurement of time at two locations many hundreds of miles apart, at a time when even the fastest message over such a distance took days. However, Eratosthenes calculated the circumference of the Earth from his home in northern Egypt without traveling anywhere.

Position of the Sun

He knew there was a way of finding a place on Earth where the Sun was directly overhead at noon—and that method involved looking down a well! Only when the Sun is directly overhead will your head cast a shadow down a well, and your reflection will be spectacularly easy to see since the very brightest part of the sky will be reflected around it. It is noon at that time at every place directly north or south of the well.

North or south of the tropics, the Sun never shines straight down a well.

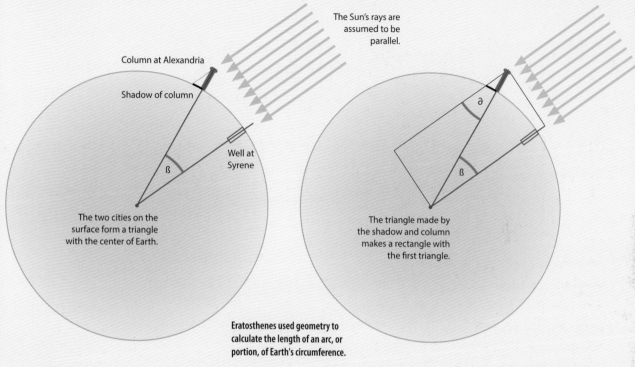

The Sun's rays are assumed to be parallel.

Column at Alexandria

Shadow of column

Well at Syrene

ß

The two cities on the surface form a triangle with the center of Earth.

∂

ß

The triangle made by the shadow and column makes a rectangle with the first triangle.

Eratosthenes used geometry to calculate the length of an arc, or portion, of Earth's circumference.

Two cities

Eratosthenes knew about a well in the town of Syrene (now known as Aswan) where this happened on midsummer's day. Eratosthenes never left his home in Alexandria, and he asked merchants who ran camel trains to Syrene how long it took to get there. That told him the distance was 5,000 stadium lengths, or stadia. At noon, he measured the length of the shadow of a vertical column in Alexandria, called a gnomon (see page 24), and then he found his answer, using triangles. Eratosthenes measured the angle (∂) that the column made with the shadow it cast, and found it was 7°. This allowed him to find the angle ß, which represented the arc of Earth's surface between the two cities Alexandria and Syrene. Adding some lines, we get a rectangle containing two identical right-angled triangles.

So angle ß is the same as angle ∂. So, angle ß is 7°. As a fraction of a circle, ß is (7°)/(360°), which is about one-fiftieth.

Final figure

This showed that Earth's circumference is fifty times the distance from Alexandria to Syrene—250,000 stadia, or 25,000 miles, which is impressively close to the correct answer. Eratosthenes contributed to many other areas of mathematics, too, and, was in charge of the Great Library of Alexandria, the ancient world's greatest storehouse of knowledge.

SEE ALSO:
▸ Triangles and Trigonometry, page 56
▸ Flat Maps of a Round World, page 82

Archimedes Applies Geometry

ARCHIMEDES MADE BREAKTHROUGHS IN MANY AREAS OF MATHEMATICS, and seems to have loved shapes of all kinds. In fact, his last words were "Do not tread on my circles." He said that to a Roman soldier who was part of an invasion force taking over Syracuse, the city where Archimedes lived. The soldier killed the great man on the spot.

This killing occurred in 212 BCE, and it followed a long siege by the Roman armies, during which Archimedes had defended Syracuse using a variety of war machines he had invented—ship-sinking claws and scorching beams made from focused sunlight. Archimedes was related to the King of Syracuse, Heiro II. One day, Heiro asked him for help. The king suspected that a golden crown he had paid for was not pure gold and that the goldsmith had mixed in cheaper metals. Since silver is lighter than gold, Archimedes realized that a particular volume of a silver/gold mix would be lighter than the same volume of pure gold. So, if he knew the volume of the crown, he could then weigh it and compare the answer with the weight of the same volume of pure gold.

The last moments of Archimedes show a Greek-style pair of compasses (see more, page 64) next to the geometric shapes drawn in the sand.

Bath-time breakthrough

Geometry could tell Archimedes the volumes of simple shapes like cubes, but what about the complicated shape of a crown? The breakthrough for this problem came in the bath. As Archimedes lowered himself into the water, some slopped over the sides. Archimedes realized

This etching suggests that after his now-famous eureka moment, Archimedes installed crowns and a displacement apparatus in his bathroom. The discovery also led to Archimedes describing the way that objects float—or sink.

Building machines

At a time when practical tasks were carried out by slaves, mathematicians had no interest in applying their minds to such matters, but Archimedes seems to have thought differently. On a trip to Egypt, he was fascinated to find a very practical use for the helix, or three-dimensional spiral. Rotated inside a cylinder, the helix could be used to pump water. Archimedes may have introduced such pumps to Syracuse, and they are called Archimedean screws today (see box, below).

A new use for polygons

Archimedes used polygons to greatly improve the known value of π. He realized that if he drew

that if the bath was full to begin with, and he submerged himself completely, the volume of water that overflowed would be exactly the same as the volume of his own body. He could find the crown's volume by submerging it in water and measuring the displaced water. He was so excited by this discovery that Archimedes ran down the street shouting "Eureka!" ("I have found it").

ARCHIMEDEAN SCREWS

Archimedean screws have been used to pump water in many parts of the world, and are still popular for some uses today because, unlike other pumps, they can deal with lumpy liquids like sewage. Tiny versions are used in hospitals to pump blood, because they do not damage the delicate blood cells as other pumps would.

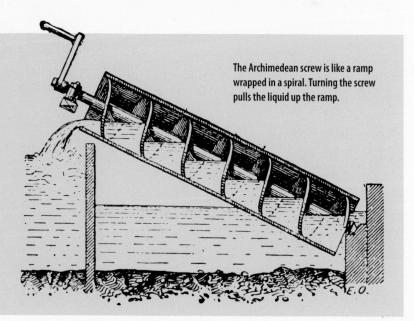

The Archimedean screw is like a ramp wrapped in a spiral. Turning the screw pulls the liquid up the ramp.

Archimedes approximated the circumference of a circle by sandwiching it between polygons. He did not need to draw these constructions. He calculated the lengths of each pair of polygons.

one polygon just inside a circle and another just outside, the length of the circle's perimeter must lie somewhere between the lengths of the perimeters of the two polygons. The more sides the polygons have, the closer they are to a circle. Archimedes went as far as an enneacontahexagon, which has 96 sides. Having worked out the perimeter length of this shape (let's call that P), he knew it was just about equal to the circle's circumference (C), and the equation for that is C = πd where d is the diameter of a circle. So, P $\approx$ πd (the symbol $\approx$ means "approximately"), and so $\pi \approx$ P/d. This technique results in the number 3.14186, which is very close to the actual value of π which is about 3.14159.

Areas under curves

Many of Archimedes' discoveries were the first versions of mathematical topics that would not be developed further for many centuries to come. Perhaps the most advanced was his approach to working out the area of part of a parabola, a kind of U-shaped curve (see page 38). The first step was to calculate the area of a triangle that would just fit into the parabola. This left two gaps, seen in yellow in diagram A on the right. He next calculated the areas of two triangles that would just fit into these yellow areas. This left four

smaller yellow areas, as shown below in diagram B. So far, this is the same kind of process that Archimedes had used to estimate the value of π. But his next step was a real masterstroke.

An infinite sum

Instead of adding four more triangles, then eight more, then sixteen more, and so on, Archimedes looked at the triangles' areas as a series of numbers (we now call it a geometric series) and considered what would happen if the series went on forever. He found the answer using a short-cut technique for adding up the series that is known today as "summing an infinite series." By doing this he was able to find an exact value for the parabola's area. This approach of dividing up areas into infinite numbers of shapes and adding their areas is the secret of integration, which is probably the most powerful of all mathematical techniques used today. However it was not developed properly until the 17th century.

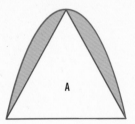

How it works

The secret of Archimedes' stomach

One of the strangest fragments of Archimedes' work refers to a children's game called the Ostomachion (which means "stomach," although no-one knows why). It involves cutting a square into 14 pieces and then reassembling them in different patterns. The fragment was badly damaged (it had been almost erased by a medieval scribe who had used the parchment to write prayers on) so it was hard to know why Archimedes would have been interested in something so simple. The probable answer only emerged this century. He had worked out the number of ways to arrange the pieces to form a square. In 2003, this number was found to be 17,152. It took four mathematicians several weeks to calculate. Archimedes was dabbling in an area of mathematics known in modern times as combinatorics.

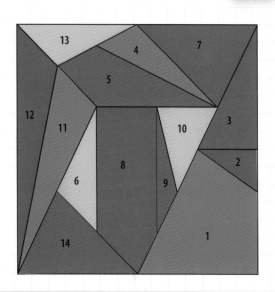

Sphere and cylinder

Archimedes' favorite discovery was that both the volume and the surface area of a sphere are two-thirds of those of a cylinder into which the sphere just fits. More than a century after Archimedes' death, his lost tomb was discovered, and on top of it were a sphere and a cylinder, carefully carved in stone.

Archimedes' tomb is once again lost. It is shown here, with its cylinder and sphere on top, in the foothills of Mount Etna, Sicily's giant volcano.

SEE ALSO:
▶ Spirals, page 22
▶ The Shapes of Perfection, page 30

Triangles and Trigonometry

HOW CAN YOU MEASURE THE DISTANCE BETWEEN STARS IN THE SKY? This was a question which Hipparchus, one of the first great astronomers, needed to answer because he wanted to map the stars.

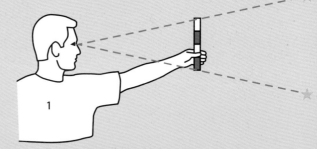

For the long-armed astronomer (1), the stars are 4 units apart, for the short armed one (2) they are only 2 units apart.

One way is to hold up a ruler (1 and 2, right), but the problem is that the answer you get depends on how long your arm is. Another method is to look along your arm, point it first at one star, then the other, and measure the angle between these two directions (3). But this is a fiddly and often inaccurate thing to do. Luckily, though, there is a simple relationship between these two methods, which Hipparchus may have been the first to explore. (He is also thought to have introduced the method of using degrees to measure angles, an idea he got from the Babylonian number system.)

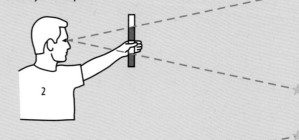

Observational equipment

To begin with, Hipparchus measured the angle between two stars. Since rods are easier to measure than arms, he probably used a cross-staff or "radius astronomical." A cross-staff makes a triangle, as

The angle between these two stars is always 25°, however long the astronomer's arms are.

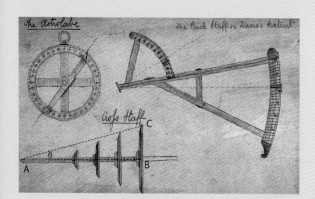

Astronomers have developed several devices for measuring the angles and positions of stars. Clockwise from top left are the astrolabe and back-staff, both of which were developed from the third device, the cross-staff. This earlier system was designed to make it easy to halve an observed angle.

AB and BC, so we use the tangent function. In the diagram left, AB (the length of the staff to the cross piece) is about 45 inches, and BC (the half-length of the cross piece) is about 20 inches. So, *tan* θ = 20/45, which is approximately 0.44.

Ancient relationship

So, we just need to find the angle θ from its tangent value 0.44. This was originally done by drawing and measuring triangles, but, later on, tables of values were written out—Hipparchus may have produced the first such table himself, but if so it has long been lost. Now, calculators, computers, and most phones have built-in tangent functions, which would tell us that the angle

shown above. The distances AB and BC are the shorter sides of a right-angled triangle, with AC being the longer, third side, the hypotenuse. The Greek letter "theta," θ, is probably used as a symbol for an angle because it looks a bit like a circle with a line through it, and angles are fractions of circles. The size of the angle θ above depends on the lengths of the sides of the triangle, and can be found using any two of those sides. It is related to them by three trigonometric functions (trigonometry is the use of the ratios between a right-angled triangle's sides and angles). They are the sine (*sin*), cosine (*cos*), and tangent (*tan*), which are defined as:

$$sin\ \theta = BC \div AC$$
$$cos\ \theta = AB \div AC$$
$$tan\ \theta = BC \div AB$$

So we can work out the angle θ just from the lengths measured by the staff. These lengths are

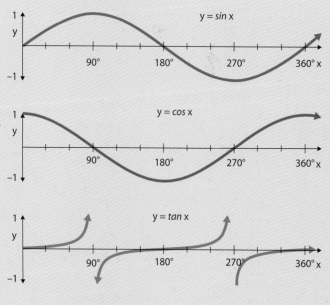

There are several ways to find the sine, cosine, or tangent of an angle. The simplest, but roughest, method is to read them off graphs like this one.

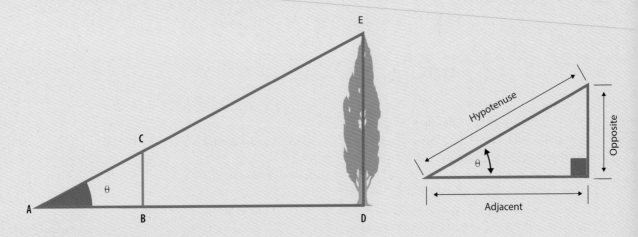

The same triangle method as used in astronomy
can be used to calculate the height of a tree.

whose tangent is 0.44 is about 24°. This value is
sometimes called the arctangent of 0.44.

Tree heights from triangles

This idea can be used to find the heights of trees,
buildings, or mountains, too. In the diagram
above, a vertical post (CB) is positioned so that
someone at point A sees the top of the tree and
the top of the post lined up. Measuring the
horizontal distance to the post, and its height,
gives us the tangent to the angle, from which we
can look up the angle itself: BC/AB is 3/4, which
is 0.75, and 0.75 is the tangent of 36.8°. There is
a second right-angled triangle here, ADE, which
includes the tree itself. Again, the tangent of our
angle equals the height of that triangle (which
is the height of the tree) divided by its distance
(which can easily be found by pacing it out—
let's say it's 100 yards). So, *tan* θ = DE/AD. This
means that DE = *tan* θ × AD. We already known
that *tan* θ is 0.75, so the tree must be 0.75 x 100
yards, which makes it 75 yards high.

General features

Rather than labeling every new triangle with
letters, we can just define the three functions in
terms of the sides of any right-angled triangle:

$$sin\ \Theta = \textbf{OPPOSITE} \div \textbf{HYPOTENUSE}$$
$$cos\ \Theta = \textbf{ADJACENT} \div \textbf{HYPOTENUSE}$$
$$tan\ \Theta = \textbf{OPPOSITE} \div \textbf{ADJACENT}$$

There are also "opposite" versions of each of the
three main trigonometric functions:

$$\textbf{secant}\ \Theta = \textbf{HYPOTENUSE} \div \textbf{OPPOSITE}$$
$$\textbf{cosecant}\ \Theta = \textbf{HYPOTENUSE} \div \textbf{ADJACENT}$$
$$\textbf{cotangent}\ \Theta = \textbf{ADJACENT} \div \textbf{OPPOSITE}$$

Triangles without right angles

All the trigonometric functions are defined by
the features of right-angled triangles, but not all
triangles are right-angled.

How it works

Algorithms

The trigonometric functions are related to each other in many ways, and the resulting theorems are used in many areas of mathematics, as well as in physics and engineering, too. When more than one angle in the same triangle is being referred to, A, B, and C are usually used for the angles rather than θ. Sometimes, x is used in place of θ. One of the most useful trigonometric theorems is $(\sin x)^2 + (\cos x)^2 = 1$. This is more usually written as $\sin^2 x + \cos^2 x = 1$. There are many other theorems too, but most fall into just three patterns:

1. Theorems about sums of angles

$$\sin(A+B) = \sin A \times \cos B + \cos A \times \sin B$$

and the similar versions:

$$\sin(A−B) = \sin A \times \cos B − \cos A \times \sin B$$

$$\cos(A+B) = \cos A \times \cos B − \sin A \times \sin B$$

$$\cos(A−B) = \cos A \times \cos B + \sin A \times \sin B$$

$$\tan(A+B) = (\tan A + \tan B) \div (1 − \tan A \times \tan B)$$

$$\tan(A−B) = (\tan A − \tan B) \div (1 + \tan A \times \tan B)$$

2. Theorems about angles of sums

$$\sin A + \sin B = 2\sin((A+B)/2) \times \cos((A-B)/2)$$

and

$$\sin A − \sin B = 2\cos((A+B)/2) \times \sin((A-B)/2)$$

$$\cos A + \cos B = 2\cos((A+B)/2) \times \cos((A-B)/2)$$

$$\cos A − \cos B = −2\sin((A+B)/2) \times \sin((A-B)/2)$$

3. Theorems about double angles

$$\sin 2A = 2\sin A \times \cos A$$

and

$$\cos 2A = \cos^2 A − \sin^2 A = 2\cos^2 A − 1 = 1 − \sin^2 A$$

$$\tan 2A = 2\tan A/(1 − \tan^2 A)$$

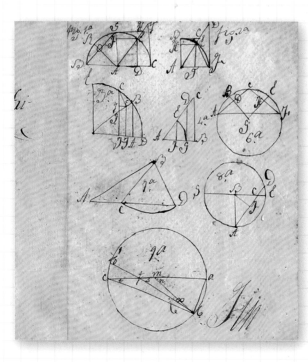

All triangles belong to one of three types:

EQUILATERAL: ALL SIDES THE SAME LENGTH
ISOSCELES: TWO SIDES THE SAME LENGTH
SCALENE: NO SIDES THE SAME LENGTH

Some trigonometric theorems work for every triangle thanks to an extension of the Pythagoras theorem. This says that for any right-angled triangle (such as the one shown on the right, with sides a, b, and c and angles A, B, and C), the relationship $c^2 = a^2 + b^2$ is always true.

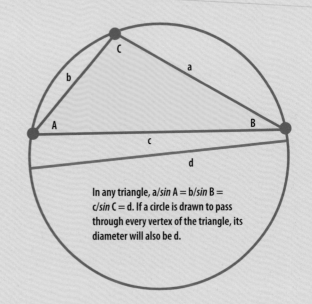

In any triangle, a/*sin* A = b/*sin* B = c/*sin* C = d. If a circle is drawn to pass through every vertex of the triangle, its diameter will also be d.

MYSTERY TRIANGLE

Triangles A and B here are each made up of the same four pieces, but arranged differently. But the area of triangle B is larger (by one square) than triangle A. How is this possible? This is a trick question. Although the human eye cannot see it easily, each big triangles' hypotenuse is not straight, but slightly bent—they are not triangles at all. This is why we can rearrange them in this confusing way.

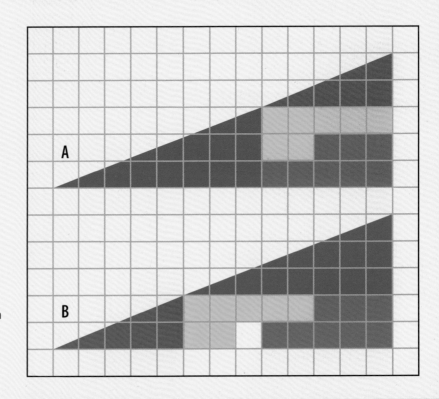

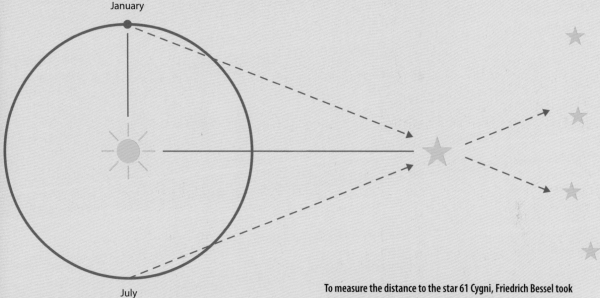

January

July

Cosine rule

The extension of the Pythagoras theorem says that for any triangle $c^2 = a^2 + b^2 - 2ab \times cos\ C$. This extension theorem still works for right-angled triangles because, when C is a right angle, $cos\ C$ = zero, so the $-2ab \times cos\ C$ part of the equation all becomes zero, and we are just left with the standard Pythagoras theorem again. The Pythagoras extension theorem is usually called the cosine rule.

Circumscribing a triangle

In the 13th century Nasir al-Din al-Tusi, a Persian mathematician who is regarded by some as the founder of the field of trigonometry, was able to prove a further link. Dividing any side of any triangle by the sine of the opposite angle always gives the same value (see the diagram above left). Going further, the value is the diameter of

To measure the distance to the star 61 Cygni, Friedrich Bessel took measurements six months apart, because the Earth takes this long to move halfway around its orbit of the Sun. This meant his two measurement points were about 190 million miles apart. He used this as the base of his triangle, and his measurements told him the angle at the far end of the triangle, so trigonometry could tell him the triangle's length. It was a very thin triangle indeed, with the long sides about 30 billion times longer than the short one.

the circle circumscribed around the triangle (see more about circles and triangles on page 14). Trigonometry does not only tell us about the sides and angles of triangles, but areas and volumes, too. The area of any triangle, for instance, is given by the formula:

$$Area = 1/2ab \times sin\ \theta$$

where θ is the angle between any two sides.

How far are the stars?

Nearly two centuries ago, trigonometry was used to find the distance to a star beyond our solar system for the first time. In 1838, Friedrich

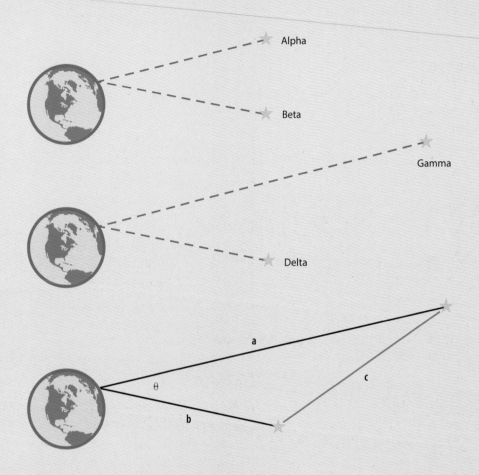

Alpha

Beta

Gamma

Delta

a

θ

b

c

Because Hipparchus could only measure the angle between stars, he could not know their distances from each other: from Earth, Alpha and Beta look just as far apart as Gamma and Delta. Only when Bessel and other astronomers had measured the distances a and b to two stars could their distance c from each other be found. Using θ, a, and b, we can work out c^2 from the cosine rule, $c^2 = a^2 + b^2 - 2ab \times \cos θ$. Then we find the square root to get c.

Bessel measured the distance from Earth to a star called 61 Cygni as about 10.3 light-years (about 60,000 billion miles). This is quite close to the actual value of about 11.4 light-years. Once the distances to several stars had been measured in this way, astronomers were able to resolve a challenge that had existed ever since Hipparchus made his measurements. He could measure the distances between stars in the sky as they appeared to him. But because he didn't know how far away any of them were from Earth, he had no way of knowing how far they were from each other. Two stars which look close together in the sky may be close together in space—but they may just happen to lie in roughly the same direction and actually be very far apart.

Beyond triangles

Trigonometric functions can be explored with no reference to triangles, and in about 1400 an Indian mathematician called Madhava found a way to define them purely through a series of numbers. Other than his many mathematical discoveries, we know almost nothing about Madhava, neither the date of his birth nor of his death—and even the location of Sangamagrama, the town where he was

TRIANGLES OF FORCE

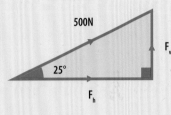

To move a sled, it's easiest to pull from as low an angle as possible. Trigonometry allows us to work out exactly how much easier. Force is measured in newtons (N). To lift an apple takes about one newton. Let's say the ox in the picture has the strength to pull with a force of 500N. To apply all this force to pull the sled would be very tricky, because to pull the sled in the direction it will move would mean lying on the ice. Maybe the smallest convenient angle he can pull from is 25°. How hard is the sled pulled along the ice in this case?

In the diagram, this is labeled as F_h (h for horizontal), and it is given by the equation $F_h = 500N \times cos\ 25°$. The cosine of 25° is about 0.9063, so the horizontal force is about 453.2N. The vertical force is given by $F_v = 50N \times sin$ 25°. The sine of 25° is about 0.4226, so this force is about 211.3N. We can check these answers by the Pythagoras theorem, $a^2 + b^2 = c^2$. If we put the values of the forces of the triangle into this equation, we get $453.2^2 + 211.3^2 = 500^2$. Working this out gives about 205,040 + 44,600 = 250,000 and the square root of that result is 500.

said to have lived, is uncertain. Madhava's series for the sine function is:

$$sin\ x = x − x^3/(3×2×1) + x^5/(5×4×3×2×1) − x^7/(7×6×5×4×3×2×1) + x^9...$$

It was rediscovered in the 1670s by Isaac Newton and Gottfried Leibniz. Madhava developed similar series for cosine and arctangent, and, as well as being new ways to define these trigonometric functions and to calculate approximate values for

them, they all contain three features that would become very powerful mathematical tools: series that go on forever ("infinite series"), and series based either on powers ("power series"), or on factorials (like 5 x 4 x 3 x 2 x 1, which is "factorial 5" and abbreviated as 5!).

SEE ALSO:
▶ Geometry + Algebra, page 98
▶ The Geometry of Architecture, page 106

Three Impossible Geometric Puzzles

TODAY, WE CAN USE MANY DIFFERENT TOOLS TO MAKE CALCULATIONS AND TO STUDY MATHEMATICS, from the simplest protractor to the most powerful supercomputer. But the Greeks developed and proved all their mathematical discoveries using only two simple tools: a straight piece of wood and a compass (also called a pair of compasses), which is more like what we might call dividers today. The Greeks also usually drew their diagrams on sand or clay, and brushed them away when finished. Even the pointiest pencil would have been pointless— and besides, no one had invented them yet.

Simple and effective

The great advantage of the Greek geometric tools was that, unlike a supercomputer, there's not much to go wrong with either tool, and if something does, it can be spotted at once. The Greeks were so successful with these simple devices that Euclid was able to prove every one of his theorems using them. It seemed to the Greeks that every mathematical problem could be solved with straight edge and compass. But could it? Sometime before 100 CE, the strange story of the Oracle of Delos began to circulate.

A holy order

The island of Delos had been struck by a deadly plague, and the Delians asked their oracle, whose job it was to exchange messages with the gods, for help. She agreed, on condition the Delians built her a new altar, exactly twice as big as her current one. The current altar was a cube, and the people of Delos were no doubt relieved to be given such a straightforward task. They built an altar twice as high, twice as long, and twice as wide, but the plague continued, as deadly as before. As with any mathematical problem, it's vital to be sure you understand the question before trying to answer it: the Delians should have asked what the oracle meant by "twice as big." What she required was a cube twice the volume of the original altar. The Delians had provided a cube eight times the volume of the original. And the oracle's apparently simple request turned out to be impossible.

Doubling the cube

To see why it can't be done, we need to use mathematics which the Delians did not have, and which was not developed for another two

Left: The remains of Delos today.

Right: Doubling the length, width, and height of the cube-shaped altar results in a an edifice eight times as big.

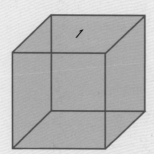

thousand years. Since it is a cube, the volume (V) of the new altar must be $V = H^3$, where H is the height (or width or length) of the cube. The Delians needed to know H when $V = 2 \times v$, where v is the volume of the existing altar. Since $v = h^3$, that means $V = 2 \times h^3$, so $H^3 = 2 \times h^3$. Therefore $H = \sqrt[3]{(2 \times h^3)}$, so $H = \sqrt[3]{2} \times h$. So all we must do is find $\sqrt[3]{2}$, the cube root of 2.

Unknown number

But there is no way to do this, neither with straight edge and compass, nor in any other way. Even the most brilliant mathematician or the most powerful computer could never find the answer. If you put $\sqrt[3]{2}$ into a calculator, spreadsheet, or computer, you will probably get an answer like 1.259921. But the cube of 1.259921 is 1.999999762, not 2. If your calculator actually gives you a 2, it's wrong—it is rounding the answer because it has no space to

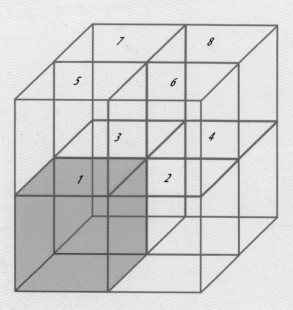

KEEP CALM AND CALCULATE

Another version of the Delian problem (told by the writer Plutarch in about 100 CE) tells us that the Delians appealed to the oracle when political infighting threatened to bring down Delian society. It was the great philosopher and fairly great mathematician Plato who helped them get started on the problem, thinking that concentrating their minds on geometry would calm everyone down. Plato was firmly of the opinion that hard thinking could solve any problem, so this is just the kind of advice he always liked to give.

Plato with his students at his school in the park known as the Akademia.

Boethius is remembered as a medieval philosopher, not as a mathematician.

irrational numbers and mathematicians did not learn how to deal with them properly until around 1000 CE, when they were studied in both Egypt and Persia.

A problem with pi

The Delian problem is one of three that vexed the Greeks and many other mathematicians who followed them. The second is, what is the length of the sides of a square with the same area as a circle with a radius of 1 unit? (The unit can be inches, meters, or any other measure). A circle's area is given by the formula $A = \pi r^2$. Here, r = 1, so the area A is π. So, the square we want, which must also have an area of π, must have sides which are $\sqrt{\pi}$ long. And, just as with $\sqrt[3]{2}$, no one can ever find out what number $\sqrt{\pi}$ is. After many years of trying to "square the circle," in about 500 CE, a

display it in full. We can approximate $\sqrt[3]{2}$ very closely, but without an exact answer, a cube with exactly twice the volume can never be found. There are many other numbers like $\sqrt[3]{2}$ that cannot be calculated exactly; they are called

TRANSCENDENTAL PI

Like $^3\sqrt{2}$, $\sqrt{\pi}$ and π are irrational numbers, but that's not all. Although we can't calculate an exact value for $^3\sqrt{2}$, we can use it in equations with exact answers, like $(^3\sqrt{2})^3 - 2 = 0$. But π is not like this, and cannot be used in such equations (to be precise, π is not the solution of any equation which uses powers and which contains no irrational numbers). π is known as a transcendental number, meaning literally that it "goes beyond" the rest. It is now known that most numbers are transcendental, but very few have been discovered, because it is so hard to prove that a number cannot be the answer to any equation.

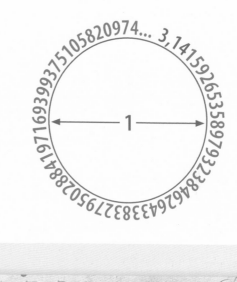

Roman scholar called Boethius claimed he knew how to do it, but unfortunately added that it would take too long to explain. This not very convincing story generated new interest in circle-squaring, and every so often people claimed to have done it. By 1775 the Paris Academy of Sciences was receiving so many "proofs" of how to square the circle, all of which it had to painstakingly check, that it refused to accept any more.

A complicated simplification

In 1882, it was proved that π is a transcendental number (see box, above), and therefore the problem could never be solved. But that didn't stop a U.S. mathematician called Edwin J. Goodwin who found a very simple, but unfortunately very daft, solution. He simply

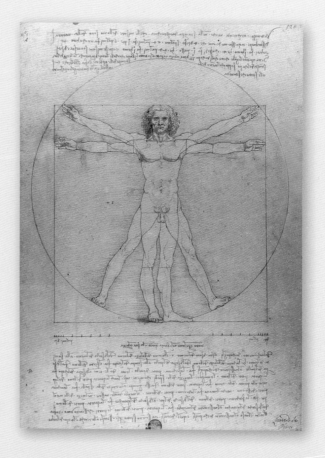

Vitruvean Man by Leonardo da Vinci gives a nod to the problem of turning a square into a circle.

forbade π to be transcendental, by defining it as 3.2. Despite the fact that this is just as possible and wise as working out the exact area of the United States by pretending it's a triangle, he managed to get this proposal seriously debated by the Indiana General Assembly in 1897, where it could even have become law if the mathematician Professor Clarence Abiathar Waldo had not been available to point out the flaw.

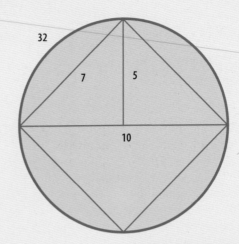

Goodwin's phony value of π allowed for these values in a circle and square—which would be impossible to draw.

How it works

Trisecting the right angle

To trisect the right angle (which, in degrees, is 90) at point A, we draw a circle of any convenient size around it. We mark the point at which this circle cuts the horizontal line as B, and draw a second circle, the same size as the first, around this point. Now we draw straight lines between A, B, and the point where the two circles cross (C). This gives us an equilateral triangle, and we know that all the angles in an equilateral are 60°. So, the angle labeled θ is 60°. The difference between this angle and our original right angle is 90° − 60° = 30°. 30 is one-third of 90, so this is the angle we wanted (labeled as α).

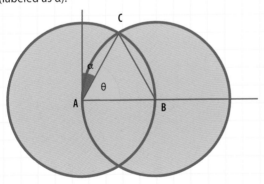

The final problem

The third problem is to divide any angle by three, and it can be solved for many angles by using a straight edge and compass (see box, left), but not for all angles. The problem can be solved for any angle using other techniques (see box, right). Failing to solve these three unsolvable problems was not a waste of time, however, because finding out why they were unsolvable led to a

In 1897, the Indiana General Assembly almost declared the legal value of π to be 3.2.

Indiana State Capitol, Indianapolis, Indiana

How it works

Trisecting by cheating

By using something more than a straight edge and compass, any angle can easily be trisected. This is how Archimedes did it.

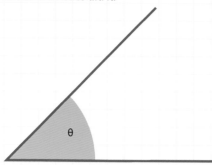

To trisect this angle, θ, we draw a circle around it (of any convenient size), and extend the horizontal line (green) across the middle of the circle.

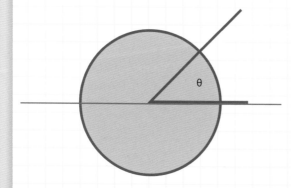

The three lines inside the circle are all radii; we copy one of them onto our straight edge.

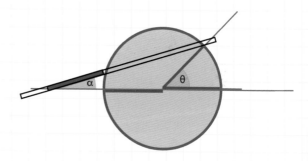

We place this straight edge onto our diagram, so that one end of the marked-off section is on the horizontal line, the other end on the edge of the circle, and the straight edge continues on passing through the point where the original sloping line meets the circle.

Positioning the straight edge in this way requires a little of skill and judgment but no mathematical operations, and this is the "cheat" in the method.

The angle α is one-third of θ, which has therefore been trisected.

far deeper understanding of the many kinds of numbers there are. Mathematics is as much about exploring the unknown as it is about carrying out useful tasks.

SEE ALSO:
▶ Euclid's Revolution, page 44
▶ Triangles and Trigonometry, page 56

Tiling and Tessellations

PAPPUS LIVED IN ALEXANDRIA, EGYPT, IN THE EARLY 3RD CENTURY, AND HE WAS THE LAST GREAT MATHEMATICIAN OF THE ANCIENT GREEK WORLD. His writings tell us about other mathematicians of his time, including Pandrosion, the first female mathematician we know of, but unfortunately, he doesn't tell us about her discoveries and only complains about some of her more useless students. Actually, Pappus complains quite a lot in his writings, mainly about the poor state of geometry in Alexandria. He did his best to improve matters, though, by developing several new theories and methods, including a new way of trisecting any angle using a hyperbola. However his method broke the rules of Greek geometry because it makes use of more than just a straight edge and compass (see more, page 52).

Pappus also studied the mathematics of tiling or connecting shapes in patterns, which is called tessellation by mathematicians. He may have been the first to prove that the only regular polygons (those with equal sides) that "tessellate," that is, cover a flat surface without leaving gaps, are the square, equilateral triangle, and hexagon.

Arabian influence

After the time of Pappus, mathematics went into a long decline in Western Europe, but developed rapidly in the Middle East. Arab and Persian scholars read the contents of ancient Greek manuscripts and started where they had left off. The mathematics of tessellation was of special interest, because Islam, the dominant religion in this region, forbids any images of living things to be used in places of worship. As a result, intricate

Even oddly-shaped quadrilaterals like this one will tessellate.

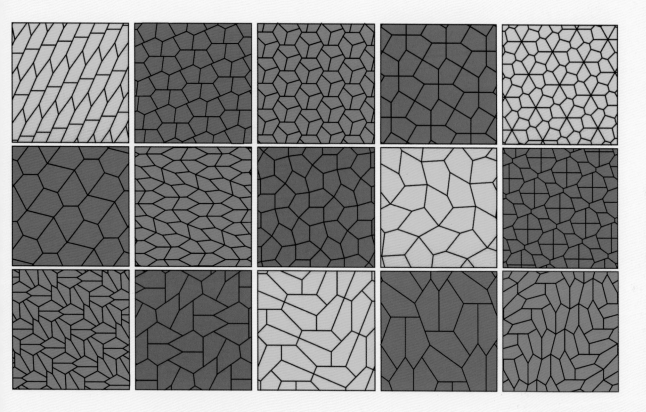

The fifteen known pentagonal tiling patterns—known so far.

and beautiful tiling patterns were used instead. Architects and geometers would meet regularly to discuss the patterns, which involved a very wide range of shapes (see box on the tiling of the Alhambra Palace, page 73).

Polygonal tiling

Many irregular polygons—that is, polygons with sides of different lengths—tessellate, including all triangles and all quadrilaterals. Pentagonal tiling is more complicated because regular pentagons do not tile, but some irregular ones do. There are fifteen known ways to tile with pentagons—all shown above—and the most recent was only found in 2015.

Wallpaper groups

There is no limit to the number of tiling patterns, but they can be classified into 17 different types—often called wallpaper groups— according to what kinds of symmetry they have. Some stay the same when rotated, shifted from side to side, or reflected in a mirror (see the box, page 72). The wallpaper groups are not easy to classify, and members of the same group may appear very different. The common brickwork pattern of a wall belongs to the cmm group. It can be rotated through 180°, reflected vertically

TILING WITH SQUARES

Squares do not have to be the same size to be used in tiling, as this pattern shows. In studying tessellations, symmetry is key, and there are three main kinds of symmetry. This pattern has just one kind: if you traced it and slid your tracing sideways, up or down, the lines on the tracing would cover the originals. This is called translational symmetry. The pattern does not have rotational symmetry because no matter what angle you turn it through, it always looks different. And it does not have reflection symmetry either. It looks different when reflected in a mirror.

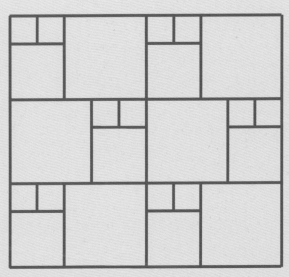

Rotation

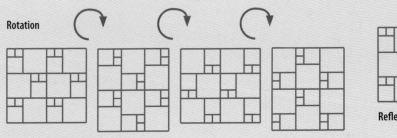

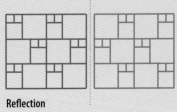

Reflection

An example of each of the 17 symmetry groups (sometimes called the wallpaper groups). The cmm type is in the middle row, second from the right.

THE ALHAMBRA

The Alhambra Palace in Grenada, Spain, was built in the 9th century. In the 13th century it was renovated and tiled in patterns that include at least 14 of the wallpaper groups. Ever since the first few groups were identified in 1948, there has been a debate over whether all 17 groups are to be found there. In 1922, the Dutch artist M. C. Escher visited the Alhambra and was so impressed by the tiling patterns that he produced many new artworks involving tessellations.

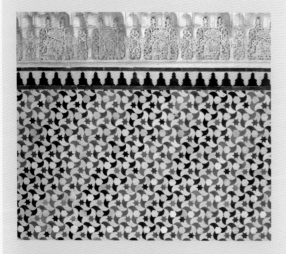

or horizontally, or slid sideways or vertically. Another member of the cmm group is shown with the full set of 17 tiling types on the page opposite. Can you spot it? It's not easy!

Aperiodic tiling

Is it possible to tile a surface in such a way that the same pattern never repeats? That would mean that if you traced out a section of the pattern, you would not find anywhere to put that tracing down where it would fit the pattern beneath. This is called non-periodic tiling, or aperiodic tiling. Given enough shapes, aperiodic tiling is easy—it is done whenever anyone lays "crazy" paving. But what if only a single shape is allowed? It was only in 1936 that the first single shape capable of aperiodic tiling was found, by Heinz Voderberg.

A Voderberg spiral.

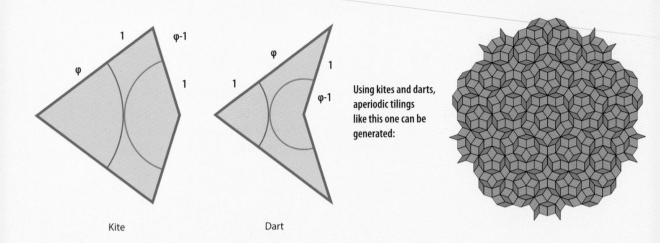

Kite   Dart

Using kites and darts, aperiodic tilings like this one can be generated:

This shape tiles in an aperiodic system into the Voderberg spiral. Voderberg was way ahead of his time; aperiodic tiling only began to be properly investigated in the late 1970s. One of the most interesting was discovered by Roger Penrose, a physicist who also worked with Stephen Hawking to develop new theories of the geometry of the Universe. Penrose is fascinated by geometry and in fact many of his breakthroughs in physics started as geometrical doodles. His new kind of tessellation is based on two shapes, the kite and the dart, both of which are constructed using the golden ratio, ϕ (see more about that on page 26).

To the edge

For a professional builder's tiler, the edge of the wall or other area to be tiled can be a problem. Usually the edges of the tiles there need to be cut off so they can fit. Mathematicians don't usually take much interest in what happens at the edges of the area to be tessellated—it is just assumed to go on forever. But Penrose tilings are an exception. Imagine drawing a line around the

outer edge of an area of Penrose tiling (shown above), then removing the tiles. Now try to re-tile within that area so that the line you have drawn once more forms the perimeter of the tiled area. You would find that only one pattern will build up to that original shape, even if it contained millions of darts and kites. This also means that, just by examining a small section of an area of Penrose tiling, it is possible to work out the shape of the edge of the whole area, no matter how huge.

Natural shapes

Penrose tilings are so clever and so strange that scientists and mathematicians alike were amazed to discover that they exist in nature, as so-called quasicrystals. These quasicrystals were studied in the 1980s, and turned out to have some odd properties. For instance, some metallic quasicrystals can transform ordinary steel into armor-plate, form super-slippery surfaces, and are good insulators of heat, despite the fact that metals are usually excellent heat conductors.

THE PENROSE TRIANGLE

In the 1950s, Penrose went to an exhibition of the artist M. C. Escher's work, and was inspired by it to invent a triangle which looks possible, but isn't. He went on to use the idea behind this triangle (now called the Penrose triangle) to invent a staircase that goes up and down at the same time. It's not only the stairs that cycle round and round; the idea did too: Escher was so impressed by Penrose's ideas that he used them in his own pictures.

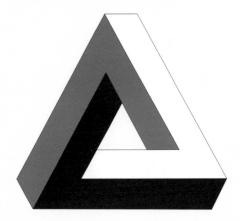

Above: A Penrose triangle.

Below: The Penrose staircase.

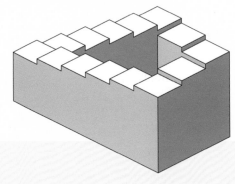

The strange and confounding geometry of M. C. Escher's pictures have inspired many artists to re-imagine everyday scenes.

SEE ALSO:
▸ The Mathematics of Beauty, page 26
▸ Crystals, page 136

Perspective

GEOMETRY OFTEN INVOLVES PARALLEL LINES, BUT THEY HAVE CAUSED PROBLEMS EVER SINCE EUCLID TRIED—AND FAILED—TO PROVE THEY EXIST. But they must do. Think about it, railroads would not work if their tracks were not parallel.

The problem is that railroad lines usually don't look parallel, unless you are hovering over them (or you are very tall indeed). And, since we live in a world of pictures, this raises a question: how is it best to draw these lines? What rules do we apply to make an accurate drawing? And how would we check its accuracy?

The Italian artist Masaccio's *Holy Trinity* painting from 1427 used a vanishing point to give the scene a true-to-life perspective.

The parallel rails appear to intersect in the distance and vanish. But no matter how we try we'll never get to that "vanishing point."

Architectural Caprice with Figures, a painting by Dutchman Hans Vredeman de Vries from 1568, has many different structures aligned to a single vanishing point.

A better view

Questions like this were of particular interest to artists, especially those working in Italy in the 15th century. During this Renaissance period, (referring to the "rebirth" of knowledge in Europe), there was new interest in learning and in the arts, and painters tried hard to produce convincing pictures based closely on the way things looked. For instance, they painted closer things larger than distant objects. Though this seems an obvious idea, many previous artists had worked to different rules, in which the most important things or people were drawn larger.

Vanishing points

Many artists experimented with ways to draw "in perspective," but the man who really changed the art world was the Italian Brunelleschi. He used a point in the middle of the image called the vanishing point. It's easy to see how this works with the railroad lines on the far left. Gradually, other artists began to use this new approach.

ILLUSIONS

Because we have all grown up with perspective drawings, we are completely used to the idea that when lines approach each other in a picture, they probably don't really. This is such a strong idea that, if we are shown a picture like this one, we assume it shows three differently sized men walking down a path with parallel sides, in front of a wall covered in lines that are parallel with each other and with the ground. Even when you are told that actually the drawings of the men are all exactly the same size, they don't look that way. You need to measure them to be sure.

Beyond the vanishing point

Though a breakthrough, a central vanishing point can't solve all the problems of perspective. What happens to railroad lines that go off in different directions? Even though no-one needed to bother much about how to draw railroad tracks in the Renaissance (iron rails were not invented until the 1760s), many scenes needed to use more than one vanishing point.

What has all this got to do with geometry?

This was just the question which fascinated Girard Desargues, a 17th-century Frenchman with very wide interests including music, drawing, stone cutting, education, botany, and mathematics. He knew all too well the practical difficulties of perspective, since, in his work as an architect, he had to draw architectural diagrams from many angles. By the 1650s, artists were comfortable with the principles of perspective, and were used to drawing such things as tabletops as irregular quadrilaterals rather than as rectangles. However, that didn't mean that artists just ignored the rules of geometry—instead they replaced them with another set of rules.

Mathematical system

The new rules weren't proved mathematically though, they were simply the ones that produced realistic-looking results. Desargues was particularly impressed that artists had learned to cope with parallel lines by having them meet at infinity, which was represented by the vanishing point. He wanted to understand the whole system of artists' rules and use them to define a new kind of "projective" geometry. He hoped that this would be of use to artists, but it seems that none of them took any notice of his work. Although some of the greatest mathematicians, including René Descartes and Blaise Pascal, were interested in Desargues's idea, most mathematicians took no notice of his writings, either. This was partly because he also attempted to improve the clarity of geometry by using new terms, which he based on botany. He called

Today, we might avoid worrying about perspective by photographing the scene we want to draw, and then tracing around the shapes in the photograph. Some Renaissance artists did something similar by using a device invented in about 1435. A series of threads was stretched over a frame to make a grid of square spaces, and the artist looked through this, copying what could be seen through each square onto a corresponding square on a grid marked on a piece of paper.

An etching by German artist Albrecht Dürer from 1525 shows the artist using a thread to map the lines of perspective coming from an object.

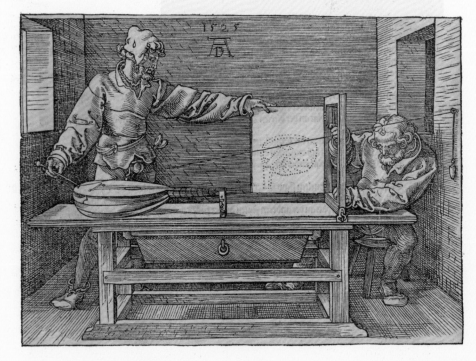

How it works

Desargues's theorem

If you had to sketch a road sign shaped as an isosceles triangle but viewed from an angle, you would know that an isosceles triangle would be the wrong shape to draw. However, what would the right shape be? In 1639 Girard Desargues came up with a theorem to give that answer.

Below: Desargues (in the center) is shown in 1643 in discussion with Pascal and Descartes (both to his left) about the nature of air. These two French geniuses were among very few mathematicians who took notice of Desargues's work on perspective.

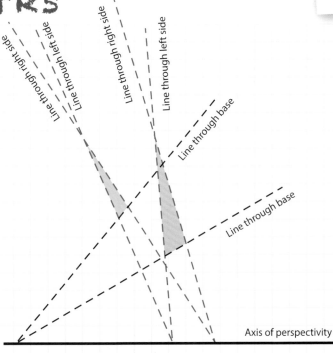

Line through right side
Line through left side
Line through right side
Line through left side
Line through base
Line through base
Axis of perspectivity

Above, Desargues's theorem is used to show that the blue triangle is an accurate perspective drawing of the green one. Lines are extended from the edges of both triangles, and marks are made at the points where corresponding lines meet (such as the lines through the bases of the triangles). If these meeting-points lie on a straight line, then the triangles are correct. The straight line is called the axis of perspectivity.

DUAL SOLIDS

Duals crop up in many areas of geometry. Take any Platonic solid, find the center of each face, and use these centers as the vertices (corners) of a new shape. This new shape will be another Platonic solid (or the same Platonic solid, in the case of the tetrahedron). These secondary shapes are duals of the first, and their definitions mirror each other: a cube has six faces and eight vertices, and its dual has eight faces and six vertices (an octahedron), and so on.

The Platonic solids and their duals.

straight lines "palms," or "trunks" if points were marked on them, or "trees" if other lines crossed them. But this idea did not catch on and made his work very hard to understand. It is only in the last two centuries that the importance of Desargues's work has been fully recognized. Today, projective geometry is essential for the development of computer games, CGI, and virtual reality.

Mathematical symmetry

Desargues's new approach also resolved an ancient mathematical problem. Mathematicians like things to be as symmetrical as possible, and so there are many pairs of theorems which mirror each other. We can work out the shape of a triangle from one side and two angles, or from two sides and one angle. Just swapping the "one" and the "two" gives both theorems. This is called duality, meaning the second theorem is the dual of the first, and vice versa (see box, left). One very basic fact in mathematics is "you can draw one straight line between two points;" that is one of Euclid's axioms of geometry. The dual of this is "you can always draw two straight lines through one point." Now, this sounds very neat, but is not always true. The only time it isn't the case is with parallel lines. However, in Desargues's new kind of space, there are no longer any exceptions. Even two parallel lines pass through one point: the vanishing point.

SEE ALSO:
▸ The Shapes of Perfection,
page 30
▸ Euclid's Revolution, page 44

Flat Maps of a Round World

PEOPLE HAVE KNOWN THE EARTH IS SPHERICAL (ROUGHLY SPEAKING) SINCE THE DAYS OF ANCIENT GREECE. However, for many centuries afterward most people only traveled short distances over land and hugged the coast during sea voyages, so the Earth's shape was of little practical importance.

Exploration begins

However, by the 13th century, some brave seafarers were voyaging around large fractions of the Earth, and making maps of what they

A map of the known world made by Martin Behaim in 1492. Christopher Columbus's voyage to the Americas that same year made the map out of date almost immediately.

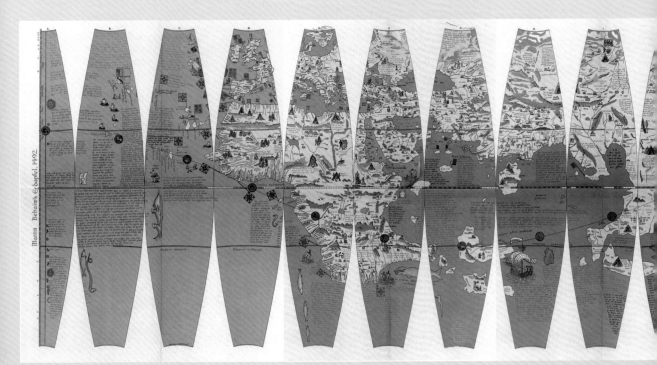

found there. By about 1500, there was enough information about the world to begin to construct a map of the entire planet. This could be done best by drawing on a globe, but carrying a globe is inconvenient for travelers. A flat map would be much better. But how to make a flat map of a solid world? This is a similar challenge to flattening an orange so that its whole outer surface can be seen. You would first have to cut the surface, and then either make a lot more cuts, or stretch the skin. And the same is true of maps of the world: they are all either full of cuts or full of stretches.

Latitude and longitude

Whether globes or sheets, all world maps (and other maps of large areas) are marked with lines of latitude and longitude. These lines are defined

Lines of latitude and longitude lie flat on the Earth, but are revealed as circular when seen from space.

Gerardus Mercator invented an early successful—but misleading—world map system in 1569.

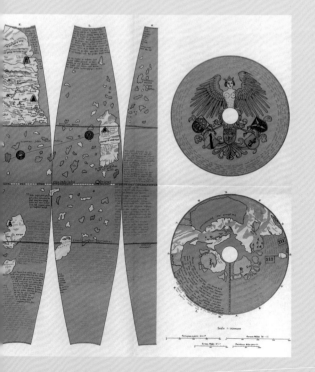

by the Sun. The Sun rises in the east and sets in the west, which defines those two directions. Lines (or "parallels") of longitude lie in east-west directions (so run north to south). When the Sun is at its highest point, the time is noon and the direction to the Sun is south for observers in the northern hemisphere and north for those in the southern hemisphere. This defines north and

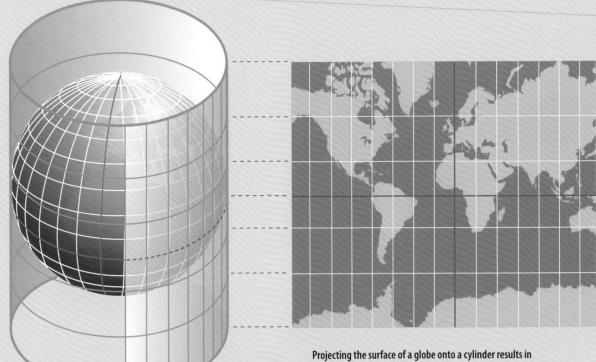

Projecting the surface of a globe onto a cylinder results in the lines of latitude getting further apart in the north and south. Mercator's projection was an approximate version of the cylindrical one. He reduced the variations between the lines of latitude a little, but his map still devotes a disproportionate amount of space to the upper latitudes, including where Europe and North America are located.

south, while lines, or "meridians," of latitude lie in north-south directions. If longitude and latitude lines were really marked on the Earth, space travelers would see them as a grid of circles.

Map projections

Different ways of mapping the Earth are called projections, and the earliest was a cylindrical projection. To see how it works and why it is called that, imagine a transparent globe of the world with a bright light in the center. Take a piece of translucent paper, like tracing paper, and make a cylinder from it that fits snugly around the globe. The inner light projects shadows of the globe's surface features, and this appears on the paper cylinder as a world map.

Mercator's map

The first good world maps used Mercator's projection, named after its Belgian inventor Gerardus Mercator, who produced the first such map in 1569. Like many who made breakthroughs in mathematics, Mercator was highly skilled in more than one subject. He was taught mathematics and geography by Gemma Frisius, one of the most learned men of his day, and was trained as an engraver by Gaspard Van der Heyden, who was the best engraver around. The three men worked together to make maps and globes, and became

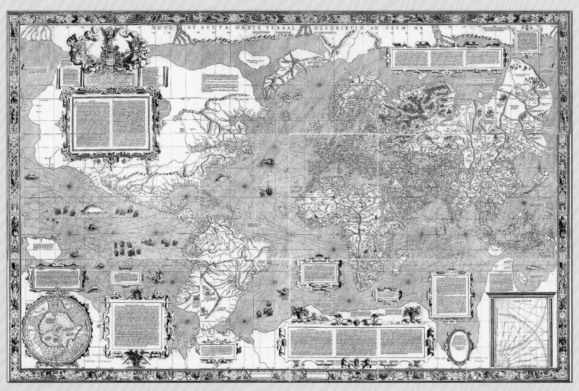

Mercator's original map of the world from 1569.

Mercator's projection distorts the size of landmasses. Each circle covers the same area on the surface of the actual globe. Northern and southern areas are stretched vertically while equatorial regions are squashed.

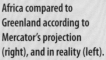

Africa compared to Greenland according to Mercator's projection (right), and in reality (left).

rich and famous. At the time, international trade was a major source of income for many countries, so good maps were very valuable indeed. In fact, pirates were as interested in stealing maps from ships as they were of plundering them for gold. However, being famous has its disadvantages. Though a religious man, Mercator had his doubts about the truth of the Bible, and in particular its account of how the world began. In 1544 he was arrested and imprisoned for months by the much-feared Inquisition, which searched out heretics.

Distorted view

Mercator's projection has been of great use to travelers for centuries and is still very popular today. One great advantage is that there are no gaps in it, and one great disadvantage is that it distorts the shapes of countries, especially close to the poles. One effect of the fact that the Mercator projection is so familiar is that people get the impression that countries far from the equator are larger than they really are. On Mercator's projection, for example, Greenland is about the size of Africa. In reality, Africa is more than 10 times larger.

Choosing a route

For centuries, the main reason for the great popularity of Mercator's projection was that it made navigation as simple as it could be, though it is quite useless for this purpose today. If you want to travel from Los Angeles to Oslo, how should you direct your course? The shortest route is a straight line, but since that would involve tunneling through the Earth, it's not ideal. An alternative is to work out the direction (that is, the compass bearing) to Oslo from Los Angeles, and go that way. The line your journey would trace on the surface of the Earth in taking this direction would appear as a straight line on Mercator's projection—and is called a rhumb line. This is, however, not the shortest route.

Great Circle

To find that, you could stick pins in Los Angeles and Oslo on a globe, and stretch a thread between them. The thread will lie along the circumference of the globe, as you could check by turning the globe until you look directly down

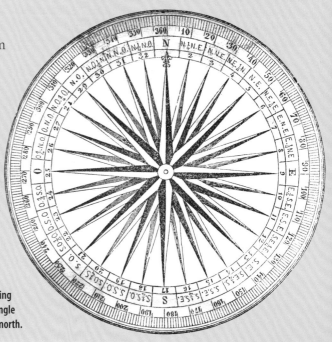

A compass bearing measures the angle from magnetic north.

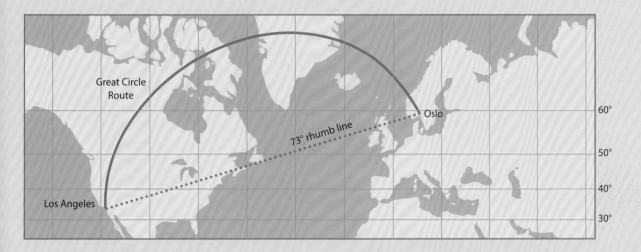

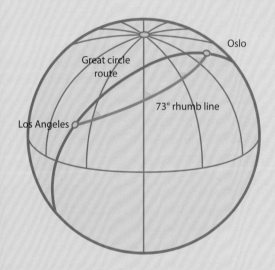

On a Mercator projection, rhumb lines are straight, and can be much shorter than great circles. To travel on a rhumb line from Los Angeles to Oslo, you just need to keep your compass bearing constantly at 73° from north. However, the shortest route is the great circle route, although the one shown above looks much longer. To follow that line, you would need to gradually change your compass bearing. Seen on a globe (as left), it is possible to see that the great circle route (red) is a little shorter than the rhumb line (green).

at the thread, which will appear straight. The thread will then appear to cross directly on top of the globe's center. But there is no way to hold a globe marked with a Los Angeles-to-Oslo rhumb line on it so that line looks straight. (There are exceptions: meridians and parallels are rhumb lines, and can also appear as straight on a globe). The line made by the thread is called a great

circle, and planes usually fly along such circles. To do this, they need to constantly change their compass bearings—a very simple task for modern computerized systems. But in Mercator's time, and for long after, constantly changing compass bearing would have been very difficult, and very likely lead to travelers getting lost. Following the same bearing for an entire journey was much easier and safer. Also, sailing ships could not be steered accurately, because of changes in wind direction, so even following a compass bearing precisely was impossible. Instead, helmsmen had to do their best to note how far from the correct bearing the wind was taking the ship, and try to make up for this by steering a little off course in

Stereographic projection of constellations in Scottish chartmaker Alexander Jamieson's 1822 *Celestial Atlas*.

A map of the Atlantic Ocean shows rhumb lines dispersing from circles. This map from 1500 is one of the first to show the Americas (just) and reveals a route to what is now Brazil.

the opposite direction when they got the chance. Trying to do this while constantly changing the target bearing would have been impossible.

Sky maps

Not all maps are of the Earth. People have been mapping the night sky for thousands of years, and they used to think that what they were doing was mapping the inside of a sphere on which the stars were stuck. Even though we now know that the stars are not all the same distance from us, the idea that the night sky is the inside of a sphere is still useful. Just like world maps, spherical star maps, though accurate, are not as convenient as flat ones. But Mercator's projection is of no use here. Most travelers do not mind that

Mercator's maps show the polar regions so poorly, but every part of the night sky can be seen from somewhere on Earth, so any star map must show every part of the sky equally well. Also, in any sky map, it is important to make sure that

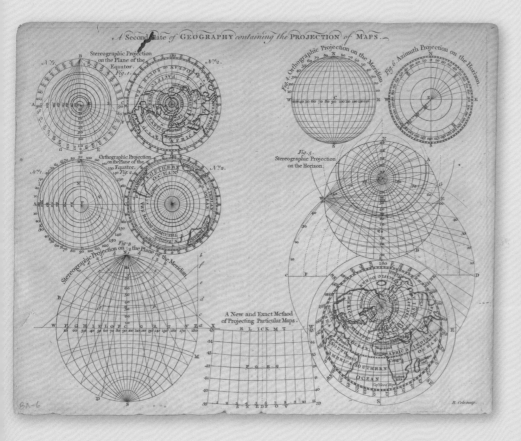

Left: A comparison of map projections was made in a 1759 book by the British mapmaker Benjamin Coles.

Below: The cylindrical projection (on the far left) and the stereographic projection (on the right) can be seen as extreme versions of the conical projection (in the middle). The red circles and dot show where the maps touch the globe.

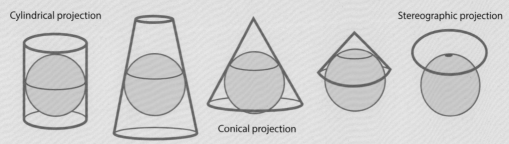

Cylindrical projection

Conical projection

Stereographic projection

the shapes of the constellations are correct. The best answer is to use a stereographic projection (see diagram, page 89). This keeps the shapes of constellations almost correct, though they do become too large near the edges. Also, only half the sky can be shown on a single map. This is no problem, of course, since that is the largest possible fraction of the sky that can be seen at any time from the surface of Earth.

Conic projection

In 1772, Swiss mathematician Johann Heinrich Lambert explored the geometry of map projections, and developed many new ones. He realized that the stereographic and Mercator projections are related by one of his inventions, the conic projection. As a cone gets flatter and flatter, its projection gets closer and closer to a stereographic one, and as it gets higher and higher it becomes more and more like a cylindrical projection.

Differential geometry

A cylinder is curved, and so is a sphere. But while we can unroll a cylinder and flatten it without any stretching, we cannot do the same to a sphere. So, we are dealing with two kinds of curvature here. The kind of curvature which a sphere has but a cylinder does not, is called Gaussian curvature after its discoverer, 19th-century German mathematician Carl Friedrich Gauss. Gauss's studies were the beginning of a new kind of geometry of curves, now called differential geometry. Gaussian curvature allows us to establish the shape of the world without

traveling far. For instance, if you lived in New York City and traveled 500 miles west, 500 miles north, 500 miles east, and 500 miles south, you would end up in the Atlantic ocean, several miles east of where you started. This is because the Earth is spherical, and your journeys have proved this fact. If the Earth was a cylinder, as some ancient Greeks believed, your trip would have brought you precisely back home.

The Remarkable Theorem

One way to find Gaussian curvatures is to use a piece of string laid in a circle. Let's say the string is 10 inches long. This is the circumference (c) of the circle it makes, and $c=2\pi r$, where r is the radius. So, $r = c \div 2\pi = 10 \div 2\pi$ inches, which is about 1.59 inches. On flat paper, the area inside the string is $A = \pi r^2$, which is about 7.95

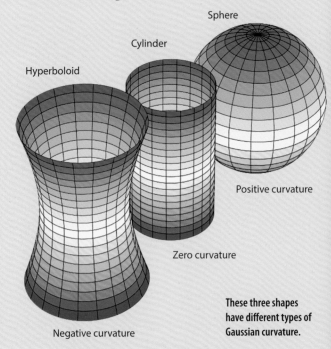

Hyperboloid

Cylinder

Sphere

Positive curvature

Zero curvature

Negative curvature

These three shapes have different types of Gaussian curvature.

Carl Friedrich Gauss was the first person to explain why map projections would always have to compromise on accuracy in some way or other.

square inches. On a sphere, which has positive Gaussian curvature, the area inside the circle will be more than 7.95 square inches, and on a shape with negative Gaussian curvature like a saddle, it will be less. Even if you change the shape of a surface by bending it, the Gaussian curvature remains unchanged: a flat piece of paper has zero Gaussian curvature, and if you make a cylinder or cone from that paper, your new shape will have zero Gaussian curvature too. Only if you stretch the paper will the curvature change. Gauss published his discovery in 1827, and it is called The Remarkable Theorem (*"theorema egregium"* in Latin) because the idea of finding the whole shape of something just by examining a tiny part of it is so remarkable. It also leads to many interesting places. For instance, why are pieces of paper flexible, but not paper spheres or donuts? Because you can flex a flat piece of paper without changing the Gaussian curvature, but not a paper sphere or torus. Why can we never find a way to make a flat map of the Earth without cutting or stretching? Because flat things and spheres have different Gaussian curvatures. The Remarkable Theorem is what lets us eat pizza without making a mess.

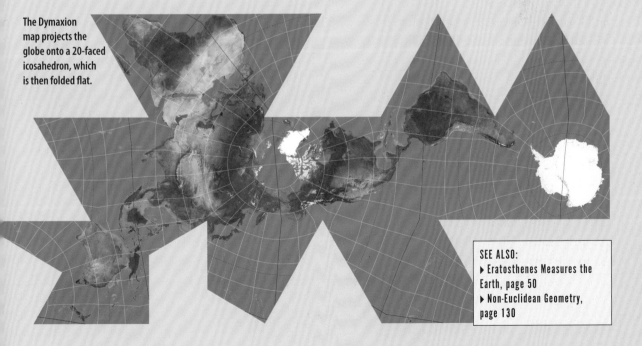

The Dymaxion map projects the globe onto a 20-faced icosahedron, which is then folded flat.

SEE ALSO:
▶ Eratosthenes Measures the Earth, page 50
▶ Non-Euclidean Geometry, page 130

Filling Space

THOMAS HARRIOT MAY BE THE MOST
UNAPPRECIATED OF ALL GREAT MATHEMATICIANS.
He made discoveries in algebra and geometry (and
also in optics and astronomy) that were well ahead
of their time, but which remained unpublished in
their full form until 2007, four centuries after he
made them. Harriot also traveled to North America
with British explorer Sir Walter Raleigh, and he
worked hard to learn the language of the Native
Americans he met, even inventing a special way of
writing the sounds of speech to do so.

Thomas Harriot (left) wrote an account of his adventures in America, *A Briefe and True Report of the New Found Land of Virginia,* the title page of which is shown above.

Harriot was Raleigh's main assistant in setting
up a colony on Roanoke Island in 1585, off
the coast of what is now North Carolina. (The
colony mysteriously vanished about two years
later, and the only human remains ever found
were the bones of a single skeleton.) Harriot may
have been the person who introduced potatoes

Cannonballs are stacked up in pyramidal structures at the St. Louis Arsenal during the U.S. Civil War.

from America to Britain, and he was one of the first Europeans to smoke tobacco (which might explain his death from facial cancer in 1621.) Harriot also helped Raleigh to design his ships, and it was their friendship that led him to a mathematical question that is still being studied today and is of great practical importance too: what is the most efficient way to use space?

Cannonball pyramid

Harriot's starting-point was a question from Raleigh. To stack a given number of cannonballs into a pyramid, how many should one use

in the base? At the time, cannonballs were usually stacked in square-based or oblong-based pyramids, or occasionally in triangular-based ones, so Harriot answered the question for all these cases, and answered the converse question too: how many cannonballs are there in a pyramid stacked on a base of a given width? These were simple questions for a mathematician like Harriot, but they launched him into the study of the nature of matter. Harriot wondered if he could explain objects as stacks of stuck-together atoms. He was quite right, but this was a dangerous thought at the time.

TO EXHAUSTION

Most proofs in mathematics are quite elegant, even though they may be very long and complicated, and they are achieved by experts in mathematics, sometimes working alone, sometimes in teams, sometimes amateurs, sometimes professionals. But there are some conjectures (unproved theorems) which cannot be solved by human ingenuity. In those cases "brute force" methods, using computers, can be tried. To tackle the Kepler conjecture (see opposite), an approach called the method of exhaustion was tried, which simply means that every single possible way to pack spheres was checked by a computer. This approach is not very satisfactory, though, because it is so hard to be certain that every possible arrangement actually has been tried. For this reason the "proof" of Kepler's conjecture obtained in 1998 was held to be only 99 percent certain. The 2014 version of the proof, which is accepted as 100 percent certain, was also based on computer analysis.

Troublesome thoughts

Everyone in Europe was expected to be a devout Christian, and there were many areas of thought which were regarded as anti-religious, including the idea that things were made of atoms. This idea of atoms had begun with the ancient Greeks, who enjoyed more freedom of thought and speech than did Harriot. The Greek atomists

believed that the world could be explained as if it were a huge machine made of atoms, with no need for a god to make it work or keep it going. However, in Harriot's time, anyone who believed in atoms could be suspected of not believing in God, and few crimes were worse than that. Nevertheless, Harriot passed on some of his ideas to the great German scientist and mathematician Johannes Kepler, including his thoughts about the packing of spheres.

Johannes Kepler was one of the first researchers to apply mathematics to natural phenomena.

Bees build a honeycomb of hexagonal cells. The cells store nectar and pollen, which is converted into honey by the workers. The honey is fed to young bee larvae which are raised in other cells until they reach adulthood.

Troublesome friends

Harriot's beliefs, and also his friends, were to lead him into danger later. First, Raleigh fell under suspicion of planning the assassination of the new king, James I, and then another friend of Harriot was arrested because he was linked with at least one of the members of the "Gunpowder Plot" which tried to blow up the English Houses of Parliament in 1605. Harriot was arrested and questioned both times but managed to talk his way out of trouble.

Kepler's conjecture

Meanwhile, Kepler decided that Harriot's cannonballs provided an example of the best possible way that spheres could be packed. That is, there could be no better way to arrange them so as to fit more of them in a box. This seemingly simple and obvious idea, which became known as the Kepler conjecture, turned out to be incredibly hard to prove. It was only solved in 1998, by the brilliant Dutch mathematician Thomas Hales, and even he needed a lot of help from a computer (see box, opposite). Still, his result was not quite certain, and only in 2014 was a complete version of the proof achieved.

Do bees know best?

A similar kind of problem to the sphere-packing one asks: if you want to build a set of identical spaces, each of a given volume and depth, what shape should they be to use the minimum amount of material? One obvious solution is to use cuboids. But in fact, trial and error will confirm that a more efficient use of material is to make hexagonal ones. Just like the Kepler conjecture, this seems simple enough, but it was only proved in 1999, again by Thomas Hales. Presumably, this is why bees build honeycombs with hexagonal cells; it is the most efficient use of their wax. (In the 4th century, Greek mathematician Pappus had suggested that "Bees … by virtue of a certain geometrical forethought, knew that the hexagon is greater than the square and the triangle and

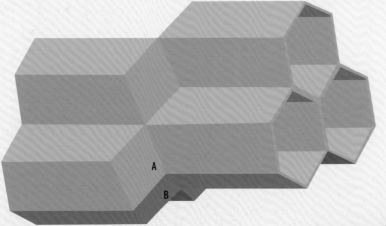

Left: The shape of natural honeycomb was found to be ever so slightly inefficient.

Below: "Kelvin foam" uses less material than a honeycomb because it has no openings.

will hold more honey for the same expenditure of material." He offered no evidence and was just following a mathematical hunch.) Nevertheless, the wisdom of the bees was only challenged in 1953 by László Fejes Tóth, a Hungarian mathematician. Bees build their hexagonal cells in paired layers, back to back. The closed ends of the cells are made of rhombuses, two of which are marked as A and B above. Tóth showed that if the bees terminated each cell with a pair of hexagons and a pair of squares instead, they would use less wax. However, it may be that bees are wiser, after all. It seems that their rhombus-shaped bases may be stronger than Tóth's alterative. If so, then a Tóth honeycomb would need to include thicker backs to the cells, which would probably use more wax after all. However, no one has yet proved or disproved this idea.

Foams of closed boxes

Mathematicians like to explore beyond the horizon by looking for the next question. In the case of the honeybees the next question is: what

if you don't need to put things into the cells or take things out? That is, if you wanted to build a set of closed boxes which occupied as much space as possible for a given amount of material, what shape would those boxes be? As in the case of the honeycomb problem, the simplest answer seems to be cubical boxes, but again, this is quite wasteful of material. In 1887, the Irish-Scottish mathematician Lord Kelvin thought he had found

The Beijing Olympics aquatics center is based on Weaire-Phelan foam.

the answer by using a shape like an octahedron with the corners cut off and with curving sides, today called a Kelvin cell. The resulting arrangement is known as "Kelvin foam."

The best yet

Nevertheless, proving Kelvin's idea (which became known as the Kelvin conjecture) defeated mathematicians and was finally disproved by two Irish geometers called Denis Weaire and Robert Phelan. They used a repeating arrangement of three different kinds of polyhedra, and found that this would use just 0.3 percent less material than Kelvin. In 2008, the aquatic center at the Beijing Olympics was constructed from Weaire-Phelan foam to celebrate the achievement. However, the story is still not finished. No one has yet been able to prove that the Weaire-Phelan solution is the best possible one.

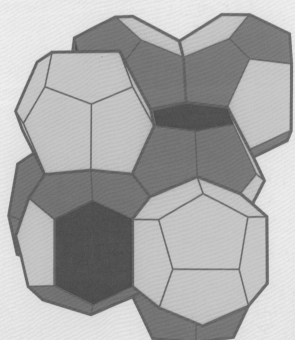

Weaire-Phelan foam is the most efficient space filler yet known.

SEE ALSO:
▶ The Shapes of Perfection, page 30
▶ Crystals, page 136

Geometry + Algebra

ALTHOUGH THE ANCIENT GREEKS PREFERRED TO USE JUST STRAIGHT EDGES AND COMPASSES TO DRAW CURVES, they knew that not all curves could be drawn that way, and it could also be very slow—drawing a parabola, for instance, required dozens of steps.

Over the years, ways were found to draw some curves by linking rods together. A marker placed on one of the rods would draw the curves when the rods were moved. In a few cases, just one rod was enough. A spiral can be constructed using a rod which is rotated around a point, while a marker moves along it. Even if you don't actually ever draw a spiral this way, it is a clear way of defining the shape. Apollonius (see page 39) wrote a whole book about this idea, but it has been lost. However, his approach was to forget about actual rods and concentrate on the idea behind them, studying the motion of

How to draw a spiral ancient-Greek style, using nothing but a straight edge.

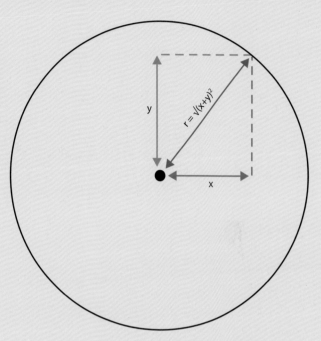

Descartes's analytical geometry can construct a circle with a radius of r using a right angle and the Pythagoras theorem.

an imaginary point as its distance and direction from a line changes. The full range of positions through which the point moves are called its locus. The idea is to define how the distance and direction change, to give you the curve you want, so, as Apollonius wrote elsewhere, "increasing the distance and angle at the same time gives a spiral."

Adding algebra

In the 16th century, the French philosopher René Descartes took this idea further using algebra (see box, right), an area of mathematics developed after the time of the ancient Greeks. Descartes realized that the concept of a locus could be used to define curves and shapes algebraically. For instance,

rather than saying that a circle is the locus of the end of a rotating rod, it can be defined as $x^2+y^2=r^2$, where r is the radius and x and y are horizontal and vertical distances from the circle's center. Together, these three values make a triangle, and by taking different values of x, a whole series of points can be defined, which mark out the circle.

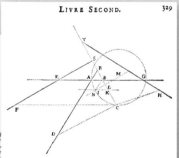

René Descartes proposed his new kind of geometry in his 1637 book *La Géométrie*.

ALGEBRA

In algebra, letters replace numbers, which allows us to grasp the idea behind number relationships. For instance, triangles with these sets of side lengths are all right-angled ones: (3,4,5), (1,1,$\sqrt{2}$), (5,12,13), (15,20,25). The Pythagoras theorem summarizes all these in a few letters and numbers, and shows us the pattern behind them: $a^2+b^2=c^2$.

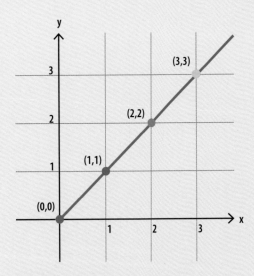

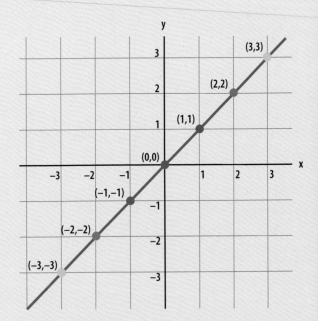

Above: The line x = y plotted on a Cartesian plane. Right: The same line with axes extended to include negative numbers.

Plotting points on a line

Descartes used any convenient points or lines to measure from, but later mathematicians soon found it easier to use two lines at right angles to each other and to label the point where they crossed as zero. The vertical line is the ordinate, the horizontal is the abscissa, and the zero-point is the origin. Distances from these lines are called Cartesian (after the Latin version of Descartes's name) coordinates. So, a straight line which starts at the origin and slopes up at 45° will pass through the Cartesian coordinates (0,0), (1,1), and so on. Because in every case the x value is the same as the y value, we can define this line as x = y. Coordinates like this make the idea of negative quantities very natural. We can clearly continue our x = y line up as far as we like. What about continuing it down? On a graph this is simple, showing that negative numbers are

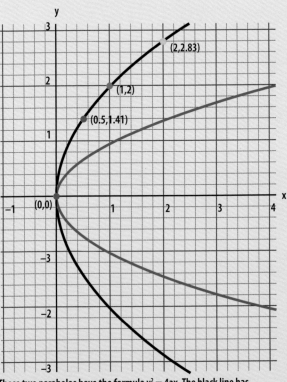

These two parabolas have the formula $y^2 = 4ax$. The black line has the constant a set as 1; in the red line, a is 0.25.

NEW NAMES

Although Descartes published his ideas in *La Géométrie* in 1637, he didn't give his new geometry a name. It is usually called analytic geometry, which refers to any geometry that involves coordinates. The area of analytic geometry which is concerned with finding solutions to equations by plotting their graphs is called algebraic geometry. Greek geometry, which uses straight edges and compasses but no coordinates, is now called synthetic geometry.

of a parabola, for example, is $y^2=4ax$, and can be used to find the coordinates of as many points as you wish, which can then be joined up to produce the parabola.

The Republic of Letters

Descartes was a keen traveler and seldom at home. Meanwhile, Pierre de Fermat, another great French mathematician of the time, rarely went out at all. However they knew about each other's work, and that of other European mathematicians, mostly through a French priest and mathematician called Marin Mersenne. He wrote letters to or met with all the great mathematicians of his day, building an international science community which became known as the Republic of Letters. It was as important then as social networks, conferences, and journals are to scientists today.

really just an extension of positive ones. This was important in an era when many mathematicians refused to use negative numbers, and some just didn't believe in them.

Analyzing lines

Descartes' new mix of algebra and geometry provided a way to apply the new math of algebra to the old math of geometry, and that meant that old geometry problems could be solved in new ways. It also made many shapes much simpler to construct. The Cartesian equation

Marin Mersenne was the founder of one of the earliest scientific academies, albeit an informal one.

Analytical geometry

Descartes used his system for just one thing. To tackle tough geometry problems by finding the algebraic equations that corresponded to them and trying to solve those equations. Although this was a powerful technique, it is a bit like using a piano to play only the tunes of songs that are hard to sing. There is much more that algebra can do. Fermat was so taken with Descartes's ideas that he used algebra to invent new shapes and curves, and found ways to graph almost any equation.

From equations to graphs

Descartes's discoveries provided a new way to solve simultaneous equations like these: $(x - 2)^2 + (y - 2)^2 = 4$ and $y = x - 2$. If we want to find the values of x and y given by these two equations, it can be done algebraically.

The blue lines show $y = x - 2$. The top red circle is $(x - 2)^2 + (y - 2)^2 = 4$ and the bottom one is $(x + 2)^2 + (y - 2)^2 = 4$.

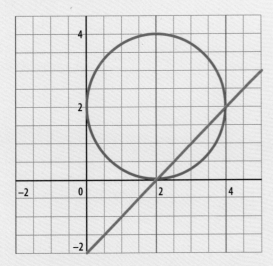

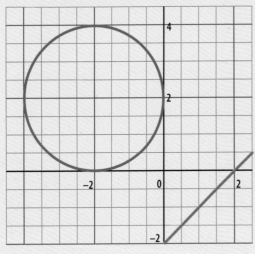

But geometry provides a much simpler method. We just have to draw a graph for each line, and we can see the answers at once: either x = 2 and y = 0, or x = 4 and y = 2. Exploring geometrical equations like this can be even more useful when a problem has no solution. For instance, what are x and y in $(x + 2)^2 + (y - 2)^2 = 4$, and y = x − 2? Plotting the lines shows they do not intersect and therefore there is no solution (unless we consider imaginary numbers, which is another topic entirely). This avoids a waste of effort in searching for answers that don't exist.

Parametric equations

If we know how to draw a curve mechanically, we can often work out the corresponding equations, specifying what the moving parts do. For instance, when a spiral is drawn by moving a marker along a rotating rod, we can see that, as time (t) passes, both the horizontal (x) and vertical (y) distances steadily increase, according to the equations $x = t \times \cos t$, $y = t \times \sin t$. So, we can use these equations to define the spiral. When a pair of equations for x and y each refer to a third quantity (which is often time) in this way, that quantity is called a parameter, and the equations are called parametric. Parametric equations are used to track the paths of actual moving objects, too. For instance, the equations for the path of a bullet or cannonball are given by

x = horizontal velocity × time

y = upward velocity × time − (16 × time)^2

The path of a projectile like a bullet can be described using a parametric equation which takes into account a third value, or parameter, in this case time alongside horizontal and vertical speed.

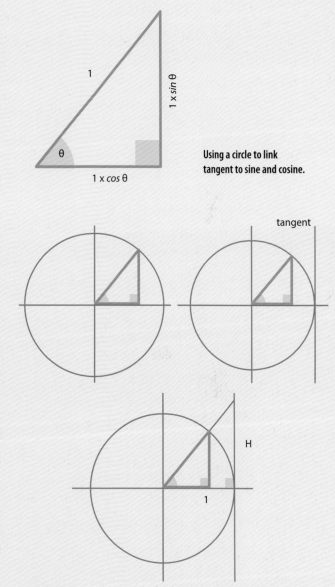

Using a circle to link tangent to sine and cosine.

The answers will only be roughly correct, because the equations don't take into account the effects of air resistance. These are the equations for answers in feet per second. (In meters per second, the 16 would be replaced by approximately 4.9; this number is defined as half of the acceleration due to the pull of the Earth's gravity). We can either use these equations to plot the whole trajectory until it lands on the ground, or to predict where the projectile will be after a certain time has passed since firing. After half a second, if the projectile was fired with an upward velocity of 300 feet per second and a horizontal velocity of 100 feet per second in an easterly direction, its position would be (very roughly) $x = 100 \times 1/2$, which is 50 feet to the right of the firing position, and $y = 300 \times 1/2 - 16 \times (1/2)^2$, which is $150 - 16 \times 1/4$, which is 146 feet above the ground.

Why "tangent"?

Why does "tangent" mean "a line that touches the edge of a circle" as well as being a trigonometric function? Although trigonometric functions are defined by right-angle triangles, these are most easily explored using circles. In any right-angled triangle, like the one above, the lengths of the shorter sides are given by hypotenuse $\times$ $sin\ \theta$ for the side opposite the angle θ and hypotenuse $\times$ $cos\ \theta$ for the adjacent, or side next to the angle.

SINE, COSINE, AND TANGENT

In any right-angled triangle, the sine of an angle is the ratio of the opposite side to the hypotenuse (opp ÷ hyp), and the cosine of that angle is the ratio of the side next to it (the "adjacent" side) to the hypotenuse (adj ÷ hyp). The tangent of the angle is the ratio of the opposite side to the adjacent one (opp ÷ adj). Because (opp ÷ hyp) ÷ (adj ÷ hyp) = (opp ÷ adj), that means that $sin\ \theta \div cos\ \theta = tan\ \theta$.

Circular link

Thanks to Descartes and the geometers that followed him, we can place this triangle on some axes and draw a circle around it (as shown on the previous page). We draw a tangent (vertical blue line) to that circle. And finally draw a second triangle using that tangent as one of its sides. The blue and green triangles have the same shape, which means that the ratios of their sides must be the same. For the blue triangle, the ratio of the hypotenuse to the base is H ÷ 1, and the same ratio for the green

triangle is $(1 \times sin\ \theta) \div (1 \times cos\ \theta)$. And, $sin\ \theta \div cos\ \theta = tan\ \theta$ (see box, left).

Families of curves

One of the many new insights that analytic geometry provided was that many curves which look different are actually closely related. This equation, for instance, generates many kinds of curve depending on the value chosen for n (a and b then fix the size and shape of the curve, and x and y are the coordinates of points on the curve):

$$(x \div a)^n + (y \div b)^n = 1$$

This was the invention of another French mathematician, Gabriel Lamé. Lamé's family needed him to work for a living, and he seemed destined for the life of a lawyer's clerk, until, in 1811, aged sixteen, he found something very surprising and wonderful in a legal library—a math book. Without telling his parents, he began to study mathematics, and for the next few years he found ways to fit mathematics into all his jobs, which included railways, building, and mining. Finally, he was able to take up a post in the French Academy of Sciences and spent the rest of his life happily exploring in detail the mathematical fixes he had developed to solve the many practical problems he had tackled. On the left, with n set to

Three $(x \div a)^n + (y \div b)^n = 1$ curves with different values of n: green = 2; red = 4, and blue = 100.

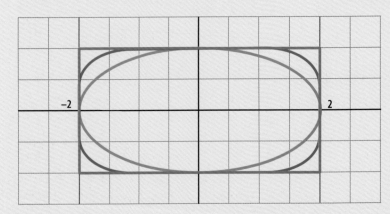

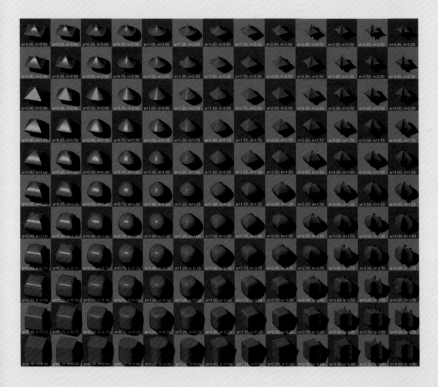

A complete family of superellipsoids with a range of values for n. The cube, cylinder, sphere, and regular octahedron are all special cases of superellipsoids.

2, the result is the green ellipse. When n = 100, we get the blue rectangle, and n = 4 gives the red shape. If n = 2.5, the resulting shape is called a superellipse, and has often been used in architecture, ceramics, and furniture-making since it was first popularized by Piet Hein, a Danish designer, in the 1960s. A three-dimensional version of the superellipse is sometimes called a superegg (partly because, unlike an ordinary egg, it can stand on either end without falling over).

The astroid

When n = 3/2, we get a curve called an astroid, which can be drawn mechanically by plotting the path of a small circle that rolls around inside a larger one. If n = 3, the curve is called the Witch of Agnesi, rudely named after Maria Gaetana Agnesi, who was the first female math professor. The curve is similar to that of low hills and water waves.

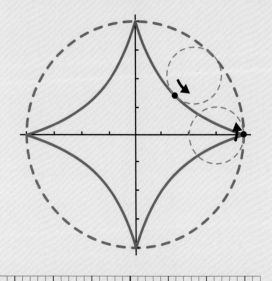

The astroid, above, and Witch of Agnesi, below.

SEE ALSO:
▶ Triangles and Trigonometry, page 56
▶ Three Impossible Geometric Puzzles, page 64

The Geometry of Architecture

FOR AS LONG AS THERE HAVE BEEN BUILDINGS, GEOMETRIC SHAPES HAVE BEEN USED TO MAKE THEM MORE BEAUTIFUL. While the golden ratio has been popular since ancient times, other shapes have been fashionable at different times and places.

The Normans of medieval Europe were very fond of pointed arches, and these were often based on equilateral triangles, but with two of the sides curved.

In 17th-century Europe, an oval formed by two circles was frequently used for laying out gardens and town "squares." One of the most impressive examples is to be found in the square of St.

The "square" at St. Peter's Basilica, Vatican City, is actually an oval made using two circles.

Right: A plate from the 1753 book *The Analysis of Beauty*, by English artist William Hogarth, contains the "lines of beauty," labeled in section 49. These curved lines were selected for their apparent beauty.

Below: An ogee arch surrounds the door of Beverley Minster in England. This arch is created by two S-like curves in mirror image of each other.

Peter's Basilica in Rome. Particular lines have also been claimed to have especially beautiful shapes, including the "line of beauty," popularized by the English artist William Hogarth in the 1750s.

The arch

Arches have been highly popular in architecture since Roman times, for a very good reason. To support a structure (whether a flat roof, a dome, a wall, or any other shape), the simplest solution is to place a beam, or lintel, on two supports, and then build on top of the beam. (For a dome, a polygon pattern of beams on columns can be used instead.) However, the lintel must bear the weight of all the masonry above it. Because it tends to curve under pressure, that means

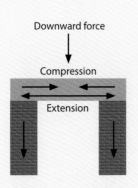

Downward force

Compression

Extension

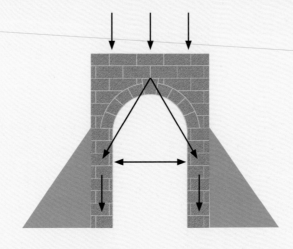

An arch redirects the downward force of the masonry so that the stones of the arch only experience a compressional (crushing) force. However, at the base of the arch there is an outward force as well as a downward one, so material must usually be added at either side of the arch to resist this sideways force.

	Force needed to crush		Force needed to pull apart	
	megapascals	pounds per square inch	megapascals	pounds per square inch
Limestone	60	8,700	2	290
Granite	130	18,850	5	725
Concrete	40	5,800	2	290
Pine	50	7,250	90	13,050
Oak	45	6,525	100	14,500

means that stone can be used. But what is the ideal shape for an arch, from an engineering point of view? No one seems to have bothered to answer this question for several centuries, during which time many different versions were tried, selected mainly because they looked nice.

that there is a strong crushing force on its top, and a strong tensional (pulling) force along its base. Stone is very weak under tension (see table, above), and, while wood is much tougher, it is hard to find tree trunks that are long, strong, and straight enough to satisfy architects with grand designs in mind.

Stronger system

The arch overcomes these problems. The shape of the arch deflects the force of the masonry above it sideways, so there is no tensional force at all, which

A dome, such as the Pantheon in Rome, is like a ring of arches, which is why domes push outward, too.

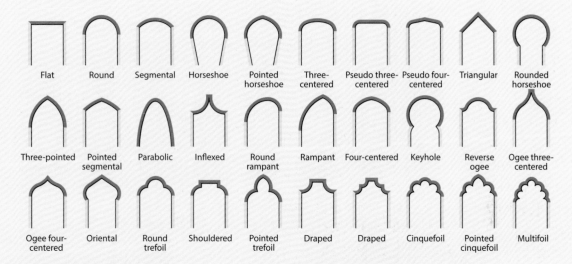

Flat	Round	Segmental	Horseshoe	Pointed horseshoe	Three-centered	Pseudo three-centered	Pseudo four-centered	Triangular	Rounded horseshoe
Three-pointed	Pointed segmental	Parabolic	Inflexed	Round rampant	Rampant	Four-centered	Keyhole	Reverse ogee	Ogee three-centered
Ogee four-centered	Oriental	Round trefoil	Shouldered	Pointed trefoil	Draped	Draped	Cinquefoil	Pointed cinquefoil	Multifoil

Architects become mathematical

While such factors as strength, wind protection, and fitting in the biggest possible windows have always been important to architects and builders, it was not until the late 17th century that anyone began to look at these issues mathematically. Before then, people just used trial and error for innovative buildings, and sometimes structures collapsed soon after being built.

A wide selection of arches have been described, and are defined by their shape, not strength.

Building by the (mathematics) book

Some of the first serious efforts to base architecture on scientific and mathematical studies were made in response to the Great Fire of London, which destroyed much of the city in 1666. The buildings and streets of the old London had evolved slowly over centuries, and the result was a very disorganized city, containing many badly built and poorly maintained buildings, all huddled close together along narrow streets. In these conditions, fire spread quickly. The new city would have to be very different, carefully planned, and with well-constructed buildings. Luckily, the 17th century was a time of great interest in

Fonthill Abbey, an immense house in southern England, was an enormous construction project. It took 17 years to build and then fell down 11 years later in 1825 because its tower was too tall.

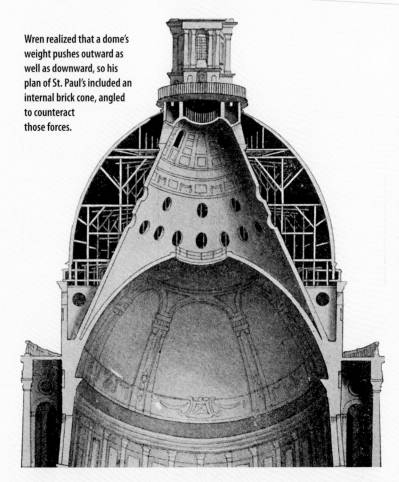

Wren realized that a dome's weight pushes outward as well as downward, so his plan of St. Paul's included an internal brick cone, angled to counteract those forces.

wanted to find the perfect arch, shaped to share the weight placed on it, so every part works equally hard. This would allow it to be as light and thin as it is possible to be. Hooke realized that this shape must be an upside down catenary, which is the curve made by a heavy rope or chain fixed at the ends.

Building with triangles

If you make a series of polygons from drinking straws taped together at their ends, you will find they all collapse very easily—except for the triangle. Crushing that requires either snapping the tape or bending the straws themselves.Because of this, the triangle is very often used in building and engineering: many bridges, for example, include triangular arrangements of girders. In the 1920s, a German engineer called Walther Bauersfeld invented a dome made of triangles, which was extremely strong and very light. Hardly anyone remembers him, but

science, and Christopher Wren, who was the greatest architect of the day, was also a scientist and mathematician. It was he who was commissioned to build the most important new building in London, St. Paul's Cathedral, and he tried to be as scientific as possible in planning it. Wren was probably encouraged to adopt a more scientific approach to building than was usual in those days by his friend Robert Hooke, who at the time was chairman of the Royal Society, an early scientific club inspired in part by the work of Marin Mersenne (see page 101). Hooke

A catenary looks a little like an upside down parabola, but is not quite the same.

How it works

The Art Gallery theorem

While most people are happy to live in cuboidal rooms, art gallery designers have more complicated shapes in mind. But, if they build a complicated shape with lots of corners for art thieves to hide behind, how many guards are needed to keep an eye on all the paintings at once? The answer turns out to be n ÷ 3 or fewer, where n is the number of corners. This relies on the very ancient discovery that any polygon can be divided into triangles (see page 31). If we divide the art gallery plan into triangles, and place a guard at one of the corners of each triangle, they can each see one whole triangle (at least). Since every part of the gallery is in one triangle or another, every part is under observation. The 1/3 arises because one out of every three corners of a triangle has a guard standing at it. It's probably fair to say that this theorem is not of much use to art gallery managers, but it has fascinated mathematicians for decades: if the guards are allowed to move around, or if there are large freestanding exhibits that block their views, or if the walls curve, the solutions become more and more complex.

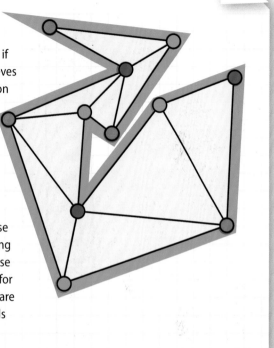

lots of people have heard of Richard Buckminster Fuller, an American engineer who popularized the design twenty years later. (Bauersfeld also invented the first modern planetarium. But it was named the Zeiss planetarium, so hardly anyone remembers him for that, either!)

Polygons are weak structures, except the triangle. The others are deformed when compressed vertically, but the triangle keeps its shape.

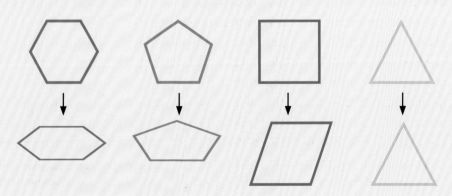

A girder bridge is a strong and inexpensive design constructed from triangular trusses.

High-tech designs

Powerful software based on the principles of perspective geometry allows a planned building to be viewed from any chosen angle. Geometry also helps to define the shapes of the building-blocks, and, for materials that need to be cut from slabs, geometrical calculations ensure that the blocks are cut out in such a way that the amount of wasted material is minimized. The cutting itself, which might be done by high-pressure water jet or laser, can also be directly controlled by the software. Buildings with unusual shapes like the Gherkin, a skyscraper in London (completed in 2010, and officially called 30 St. Mary Axe), use many geometric principles which were known for decades but could not be calculated precisely enough before sophisticated computer programs were developed.

New jobs for geometry

The geometry of the Gherkin is unusual in three ways. It has no corners, it has a central bulge, and it spirals. These three features are

PARAMETRIC MODELING

Until the late 20th century, the geometry used for building was usually based on simple shapes like cuboids (for the rooms) and triangular prisms (for the roofs), because it is so difficult to work out how more complicated shapes will function under the force of their loads and of the wind. The problem was not that the physics or geometry are unknown, but that the actual calculations were so time-consuming. This problem was made much worse by the way architects and their clients work, by constantly revising some details of the design. Because many parts of a building support or are supported by others, and because changing a single room will involve changing its neighbors too, any change will often mean almost every calculation must be redone.

Nowadays, a technique called parametric modeling makes things much easier. Once a model of a planned building has been developed, parametric modeling software automatically calculates the effects of changing one "parameter" (such as wind speed, building material, wall thickness, or ceiling height) on the rest of the design, flagging up required changes. For instance, changing the building material might require walls to be made thicker and foundations deeper (to take the extra weight), but also allow less powerful heating and air conditioning systems to be installed, and noise insulation to be reduced.

all useful as well as beautiful, and they were included partly to deal with concerns which might not have bothered architects a century or so ago. Do people near the building feel

The Montreal Biosphere, a dome based on interlinked triangles, was constructed in the Canadian city by Richard Buckminster Fuller in 1967.

physically comfortable? Do they find the building disturbing? Does the Gherkin use energy efficiently? On windy days, rectangular-based skyscrapers become very unpleasant neighbors, because the wind, being forced to change direction rapidly at the corners, forms whirlwinds there. The Gherkin's round, cornerless shape avoids this problem. The central bulge reduces wind speeds near the base too, and it also means that people nearby looking up at the Gherkin cannot see the top, which makes the building seem shorter and therefore less intimidating.

"Gherkin" is the British word for "pickle."

Less sunlight is blocked than by vertical-sided skyscrapers, as well. The overall shape of the Gherkin allows the wind to blow through when the windows are open, which saves on air conditioning bills on hot days. This effect was accomplished by turning each floor a little more than the one below, creating a spiral which funnels air round when the wind blows.

SEE ALSO:
▶ The Mathematics of Beauty, page 26
▶ Crystals, page 136

Prince Rupert's Challenge

PRINCE RUPERT WAS THE NEPHEW OF CHARLES I, KING OF ENGLAND, and a great deal of his life was spent as a soldier. First, when he was 22, Rupert fought on the side of the Royalists against the Parliamentarians in the English Civil Wars, which started in 1642. When the Royalists lost the war (and King Charles was beheaded), Rupert (who was German) moved abroad, fighting on many sides in many countries until returning to England once the monarchy was restored in 1660.

brass which looked like gold. In mathematics he became an expert in making up codes and breaking those of others, a valuable skill at a time when England was full of plots and conspiracies.

Cutting a cube

Rupert also loved geometry, and his interest led him to ponder whether a hole can be cut in a cube large enough for an identical cube to pass through

This kind of life is unusual for a scientist or mathematician, but Rupert spent the little time he wasn't fighting studying both subjects. He invented some exploding glass drops called Rupert's Drops, and a new kind of

Prince Rupert, seated left, presents the battle plan to his uncle, Charles I, on the eve of the Battle of Edgehill during the English Civil Wars. He was a ruthless soldier, but he was kind to animals, and traveled everywhere with his dogs.

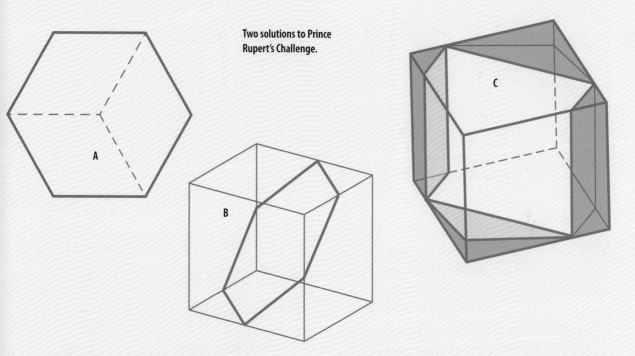

Two solutions to Prince Rupert's Challenge.

it. In 1693, Rupert bet that it could. But how? Rupert was not a good-tempered person, so he probably reacted very badly when the person who solved the challenge was John Wallis, who was not only a Parliamentarian but Rupert's personal opponent too. For, while Rupert had been making and breaking codes for the royalists, Wallis was doing just the same thing for the Parliamentarians.

The solution

If you hold up a hollow, one-inch cube so that you can just see one face, and then cut away that face and the one behind it, then that is the largest hole you can make from that point of view. It won't be quite big enough to allow another one-inch cube to pass through it. So can we make a bigger hole? John Wallis's solution was to angle the cube until we are looking straight at one of its corners

as in figure A. Its outline is then a hexagon. If we cut away all that we can of this profile without destroying the cube entirely, we will be left with a hexagonal slice, like figure B. (The sides of the original cube are just shown for comparison.) A bit of trigonometry reveals that a cube of side length 1.035 inches will just pass through this fragment. In fact, there is a better solution, found about a century after Wallis won Rupert's bet. The best hole looks like figure C, and it can accommodate a cube with a side length of about 1.06 inches.

SEE ALSO:
▸ Tiling and Tessellations, page 70
▸ Perspective, page 76
▸ Filling Space, page 92

Higher Dimensions

ONE OF THE THINGS THAT MAKES BEING A MATHEMATICIAN FUN is that you can explore other worlds that seem very different to the real one. However, sometimes these mathematical explorations reveal that our world is actually arranged in a different way than we thought. One such case is the exploration of dimensions.

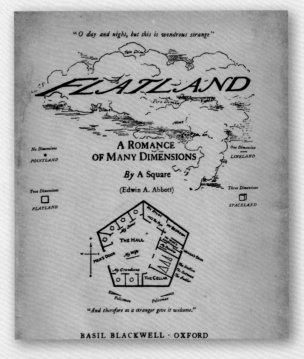

Several other-dimensional worlds are explored in *Flatland*, written in 1884 by Edwin A. Abbott (the A. is also for Abbott, oddly enough).

You are three-dimensional but these words are two-dimensional; they have height and width but no depth. Straight lines have only one dimension, and points have no dimensions at all. Imagine that you are two-dimensional, like a word on a page. You have width and height but no thickness. You cannot see out of the paper into the space above it, only sideways. You can only see the edges of any shapes near you, so your vision would be only one-dimensional.

What do we really see?

The same applies to us. We see because light makes two-dimensional patterns on the backs of our eyes (retinas). It is our brain that puts together the slightly different images from both our eyes to tell us something about the shapes of objects in our three-dimensional world.

Sometimes you need to touch an object to discover its shape. Our brain sees these shapes as faces—and faces, with their noses, lips etc, stick outward. But take a closer look; are these famous faces as they first appear? (Hint, left to right it is Albert Einstein, Nelson Mandela, and Ludwig van Beethoven.)

Together with our sense of touch, this means we can get a very clear idea of what three-dimensional objects are like, even though we can only see two-dimensional images.

A sphere visits Flatland

But what would a flat person make of three-dimensional objects? Let's push a sphere through Flatland and out the other side. A flat person would see a dot which grew into a circle before shrinking and vanishing. A cube (if it arrived by one corner first) would appear as a triangle which also grew and shrunk, and more complex solids would appear as more complicated

changing shapes. But flat people would recognize them all as three-dimensional visitors because they would all grow and then shrink, and if this happened a lot we would soon be able to recognize different kinds of shapes. What can we three-dimensional people learn from this thought experiment? Perhaps we could get used to seeing four-dimensional shapes. Depending on its orientation, a four-dimensional cube (called a 4-cube, hypercube, or tesseract) moving through our world could look like a normal cube (a 3-cube), except that it would grow and shrink before our eyes; a 4-sphere would behave as a growing-then-shrinking 3-sphere,

Edwin Abbott Abbott drew this diagram to show a three-dimensional sphere descending through Flatland. The two-dimensional inhabitants of Flatland first see a dot which grows into a circle before shrinking back to a dot again, and then vanishing.

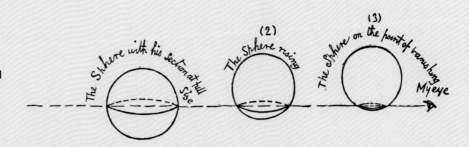

Shape	Number of			
	Dimensions (D)	Vertices (V)	Edges (E)	Faces (F)
Point	0	1	0	0
Line	1	2	1	0
Square	2	4	4	1
Cube	3	8	12	6
Tesseract	4	16	32	24
5-cube	5	32	80	80
6-cube	6	64	192	240
Formula		2^D	$D \times \frac{V}{2}$	$(2 \times F_{(D-1)}) + E_{(D-1)}$

and so on. We can also explore higher space mathematically, by thinking of spaces of different numbers of dimensions as a series, and see how the dimensions we are familiar with relate to each other, and to their number of vertices (corners), edges, and faces. We can then apply these relationships to higher dimensions, as in the table at left. A pattern emerges that lets us come up with general formulas. For example, the face formula equation means "take the number of faces of the shape in the row above, double it, then add the number of edges of that shape."

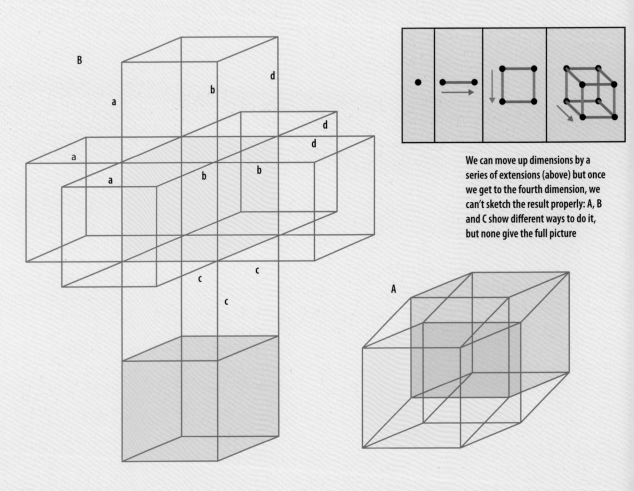

We can move up dimensions by a series of extensions (above) but once we get to the fourth dimension, we can't sketch the result properly: A, B and C show different ways to do it, but none give the full picture

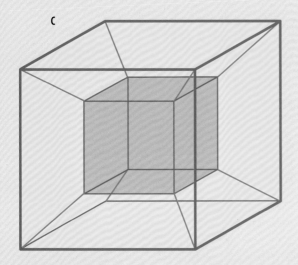

Entering hyperspace

We can think of a line as a point that is extended sideways, a square as a line that is extended upward, and a cube as a square that is extended outward (out of the page, that is). If we try continuing this series, we can say that a tesseract is a cube that is extended into the fourth dimension—entering a space, or hyperspace, beyond our senses—as in figure A on the left. Nevertheless, can we draw a picture of this extended cube? We could start by adding a new cube to each face of a normal cube as in figure B. But this is not quite right, because the outer cubes should be in contact with each other. The three lines labeled as a in figure B are really all the same line, and so are the three b, c, and d lines, and so on. If we allow the cubes to be distorted—all flat drawings of real, three-dimensional objects involve distortions—we can correct this, creating figure C above. The only other problem with this drawing is that the outer yellow cube should be the same size as the central one.

While a cube is six squares arranged in three dimensions, a tesseract is eight cubes in four dimensions. The green lines are the edges of distorted cubes in this view.

Hidden dimensions

Physicists can explore higher dimensions, too. The most recent attempt to explain the workings of matter and energy is string theory, which is based on the idea that all the particles we know of (such as electrons) are made of tiny, vibrating structures called strings. But to make this theory add up, the string vibrations must take place in a space of ten dimensions. To explain the rather obvious fact that no one has noticed this before

INSIDE A TESSERACT

If we actually came across a hollow tesseract with sides 10 feet long that weren't transparent, it would just look like a cube. But if it had doors in all its faces, exploring it would reveal its true nature. Entering any door would take us into a cubical room ten feet high with a door in each wall, a trapdoor in the floor, and another in the ceiling. In fact, it would look just the same as the inside of a cube. But go through any door, and instead of being outside again (or looking at the ground or sky), you will find yourself in an identical room. From that room, at least three of the doors will take you to another identical room, while the rest will take you outside (or show you the ground or sky). If you keep exploring you will find seven rooms, and a total of 54 doors, of which nine open to the sky, nine to the ground, and 36 take you outside again, all inside a single cube, ten feet high.

How it works

Weird dimensions

Although it is useful to explore spaces with more dimensions by learning about the ones we are familiar with, this approach does make it seem as if they are just like ours, but bigger. But this is far from the case: each has its own unique weirdnesses.

4D: Although three-dimensional space curves when gravity is present, these curves are always smooth, and a spacecraft or planet will follow them like a train following its rails round a bend. But in four-dimensional space, directions can change

instantly—which would presumably wreck any 4D objects passing by.

5D: Although there are four-dimensional versions of all five Platonic solids, in five-dimensional space there are no versions of the icosahedron or dodecahedron.

7D: In our space, we can gradually squash a sphere to make a more and more oblate spheroid. Some "exotic" spheres in seven-dimensional space can also change to spheroids—but they do so suddenly, without going through all the intervening shapes first.

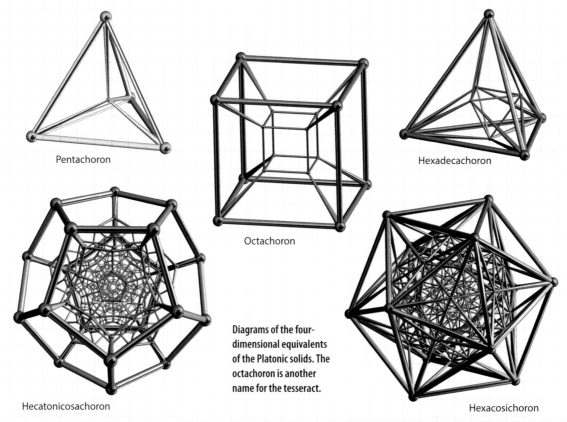

Pentachoron

Octachoron

Hexadecachoron

Hecatonicosachoron

Diagrams of the four-dimensional equivalents of the Platonic solids. The octachoron is another name for the tesseract.

Hexacosichoron

he define where anything is in this space? To define the position of an object in one-, two-, or three-dimensional space, you need one, two, or three coordinates—so, your position in space might be described as 1,000 miles west of the Greenwich meridian, 500 north of the equator, and 2,560 miles above the center of the Earth, or as the numbers (-1000, 500, 2560). An infinite-dimensional position would need an infinite number of coordinates. Is this even possible?

(a ten-dimensional teacup, for instance, would hold over 30,000 gallons of tea), physicists have two theories. One is that we are a bit like ants living between two sheets of glass. We are unaware of any further dimensions, because something blocks us from entering them. Or, the extra dimensions might be curled up ("compactified") so tightly that they are too small to explore. We know that three-dimensional spaces can curve or warp (see more on page 160), so this might be possible. This is a bit like the way that fine black threads on paper look like one-dimensional lines, but a magnifying glass would show they are actually three-dimensional solids.

David Hilbert and a curve he developed (above) which is one-dimensional, yet can fill a two-dimensional space.

Hilbert space

In fact, it is quite simple, since mathematicians are well used to dealing with infinite sequences. The simple series (1,2,3,4 ...) goes on forever, and therefore it can define such a point. So, we can define some shapes quite simply. A two-dimensional circle can be defined as a set of points all equally distant from its center, and a three-dimensional sphere has exactly the same definition, as does a 4-sphere, a 5-sphere, and so on—right up to an infinite sphere. Hilbert's space is used by physicists studying quantum theory, an area of science in which particles have an infinite number of possible states.

To infinity

In the 1900s, a German mathematician called David Hilbert began to explore a space with an infinite number of dimensions. How could

SEE ALSO:
▶ The Shapes of Perfection, page 30
▶ Geometry + Algebra, page 98

Topology

Leonhard Euler's surname rhymes with "oiler." Despite being nearly blind for much of his life, this German was one of history's most successful mathematicians.

GEOMETERS HAVE ALWAYS BEEN INTERESTED IN CLASSIFYING SHAPES, from the Platonic solids to the forms of snowflakes. But the shapes they deal with are usually much simpler than those in the real world, most of which are so irregular that they don't even have names, or if they do, those names (like "pear-shaped") don't have formulas to define them precisely. But, in 1799 a method was found to define all shapes in a new way.

The secret is not to bother with the precise shape of an object, but to choose some feature that it shares with others. This makes sense, because "pear-shaped" does not refer to a precise shape in any case, it just picks out a few features which all pears share. Even though it would be hard to list what those features are, that doesn't stop us from using the term successfully.

Hole numbers

The person who introduced feature definitions was Leonhard Euler, who classified shapes according to the number of holes they have: so, spheres, cubes, pears, and bananas are all zero-holed shapes, while tubes, donuts and hoops all have single holes. Violins have two holes, a ten-runged ladder has nine, and so on. In doing this, he invented a new area of mathematics now called topology. This system is a good

classification because it is simple, clear, and focuses on a useful feature. If you have a big, soft, beach ball which is partly deflated, you can squash it fairly easily into an oblate spheroid, or a pear, or a banana shape. But, you could not form it into a donut shape without puncturing it.

Defining features

To define the differences between shapes with different numbers of holes properly, some kind of formula was needed—and Euler had one:

Vertices – Edges + Faces = 2

This is true for any shape with no holes in it, so, for instance, a cube has 8 vertices, 12 edges and 6 faces, and 8-12+6=2. A tetrahedron has 4 vertices, 6 edges and 4 faces, and 4 - 6 + 4 = 2. Since we can reshape a cube into a tetrahedron,

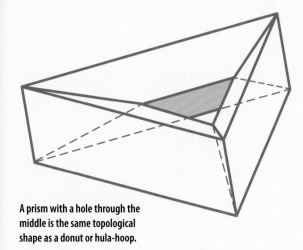

A prism with a hole through the middle is the same topological shape as a donut or hula-hoop.

sound strange, but in fact the topology of an object is important. You always want the correct number of holes in your bucket, for instance, and, for all our differences, all humans have the same topology—which we share with all other

or into a sphere, we can assume the formula works for all no-holed shapes. But the answer is different for shapes with a hole, like this triangular prism with a triangular opening in the middle. It has 9 vertices, 18 edges, and 9 faces, and 9 − 18 + 9 = 0. In fact, in any shape with a hole, the right hand side of this equation (which is called the Euler Characteristic of the shape) is zero. If there are two holes, the Euler characteristic is -2. For any number of holes, the Euler characteristic is given by

Vertices − Edges + Faces = 2 × (1 − Holes)

This is particularly useful to have as a formula since it is not always obvious that shapes are topologically the same. That is, the same in the sense of having a certain number of holes. Any shape which can be molded into another without any gluing or cutting counts as the same shape to a topologist.

Important similarities
Classifying things according to holes may

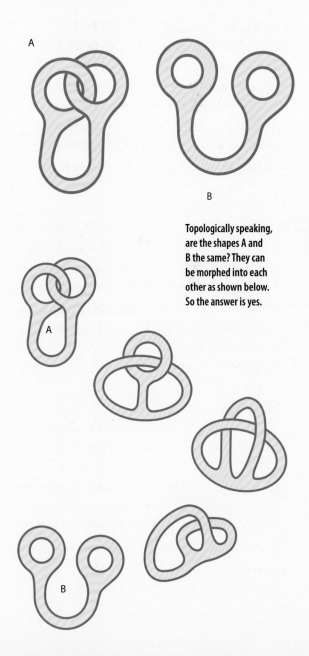

Topologically speaking, are the shapes A and B the same? They can be morphed into each other as shown below. So the answer is yes.

TOPOLOGICAL MAPS

When you are traveling underground, you don't care what twists and turns your train takes, you just need to know how to get to your destination. So, in 1931, a British engineer called Harry Beck prepared a map of the London Underground subway in which all the lines are straight and all stations are about the same distance from their neighbors (facsimile shown below). The one thing that is accurate about Beck's map is its topology—the connections between the stations. The map was so clear and useful that versions of it have been used ever since, and it is copied for subway networks in other cities (right).

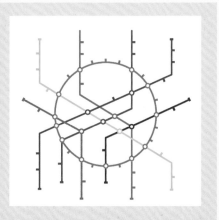

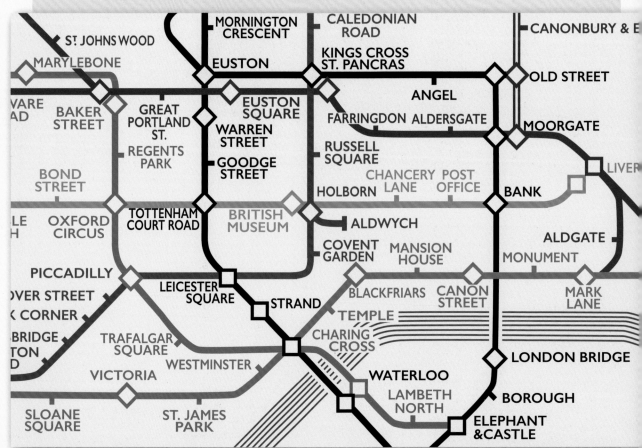

TOPOLOGY IN THE OPERATING THEATRE

Each of our internal organs varies in size and shape according to our age and state of health, and will be different from other people's. What matters is how they connect to each other, and therefore how many holes each has. That is, their topologies. If one of the four openings into the heart becomes blocked, the owner will probably die unless a surgeon either unblocks it or replaces it with a new hole, by performing a bypass operation. If someone is shot, it is the new holes that will kill them unless they can be closed again. And, when an opening (called a fistula) appears between an artery and a vein, it too must be closed again if the patient is to recover. This is not just a way of talking about surgery. X-ray images today are so complex and detailed that computers must work on them before any human can make sense of them, and in many cases the main thing the computer does is to work out what the images can reveal about the topology of the organs being studied.

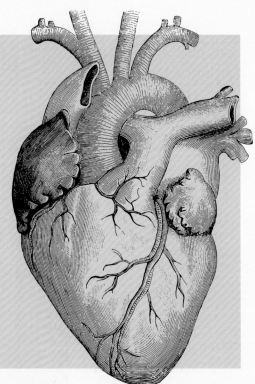

mammals too, more or less. And, for plumbers, cave-explorers, and surgeons, understanding the topologies of the systems they work with is about the most important part of their jobs. Topology is probably the most active area of mathematical research today, and has many practical implications, too.

Making connections

Topology deals with surfaces and their edges and connections, and we can use it to invent and study many new structures. The most famous is the Möbius strip. It is just a strip of paper with a twist, but unlike other pieces of paper, it only has one side. This soon becomes clear if you start coloring a section of it in. Every part of it will soon be colored, without you ever having to turn it over.

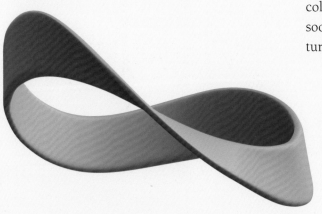

The Möbius strip, named after its German discoverer August Möbius, is a three-dimensional object with one edge and one face.

SEE ALSO:
▶ Flat Maps of a Round World, pages 82
▶ The Poincaré Conjecture, page 152

Fitting Circles into Triangles

TODAY, MATHEMATICS IS AN INTERNATIONAL SUBJECT, with mathematicians of different nations communicating by social media and gathering at conferences all over the world. In fact it is a more international subject than most, because all mathematicians speak the language of equations.

However, this is a relatively recent development. While Arabic mathematicians were closely linked with the mathematics of the Greeks, in India, China, Japan, and elsewhere, mathematics was developed independently, and techniques and theorems used in one part of the world might not be discovered elsewhere for decades or even centuries. Sometimes mathematicians unknown to each other discovered the same thing almost at the same time.

Hidden figures

For instance, in the late 18th century, Ajima Naonobu, a Japanese geometer, studied how to fit three circles into a triangle so that as little space as possible was left. However, today this is known as the Malfatti problem, since the Italian mathematician Gian Francesco Malfatti studied it just a few years later, in 1803.

Masonry problem

Malfatti viewed his problem, at least to begin with, as a very practical one. Given a piece of

Gian Francesco Malfatti was a leading geometer in Italy in the late 18th century. He helped to set up what became Italy's national academy of science. His work on the problem that bears his name was carried out in the last years of his life.

Diagram A shows Malfatti's original approach. Diagram B changes the approach to include an "incircle" which touches three sides of the triangle. Diagram C shows simplest solutions that work for thin triangles.

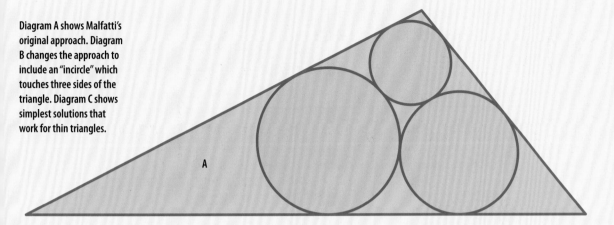

marble in the shape of a triangular prism—any shape of triangle—what are the largest three columns that can be extracted from it? Malfatti thought he had proved that the solution was a maximum of three circles, all of which must be in contact with the other two and with two sides of the surrounding triangle. This condition was soon proved to be correct by several other mathematicians in different ways.

Another problem

But math can be slippery, and although all the proofs were correct, there was a problem. The sketch that Malfatti had in mind, which seemed a very fair one, actually restricted the problem unnecessarily. Diagram A above shows that all three circles are in contact with each other, and it is true that, if this is essential, Malfatti's solution is correct. But of course, if all you want is to

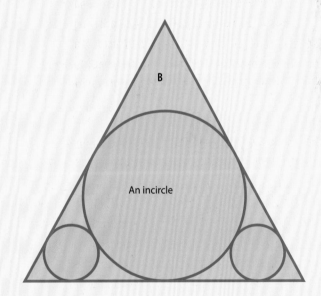

An incircle

GRAPHIC LIES

Diagrams can cause other kinds of problems, too. Sometimes, designers use perspective to mislead us. The large diagram suggests—at first glance at least—that TV Channel B has the most viewers. The smaller diagram reveals the truth.

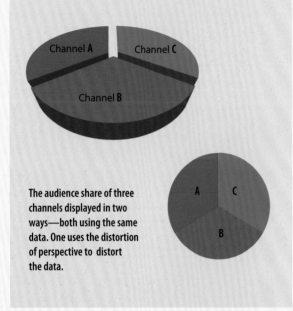

The audience share of three channels displayed in two ways—both using the same data. One uses the distortion of perspective to distort the data.

make three columns without waste there is no need for them to be in contact. If this restriction is dropped, we can then fit in one circle which contacts all three of the sides of the triangle (this is known as the incircle; see diagram B). Then, the other two columns can be squeezed in, and a little less marble is wasted. For very thin triangles there is an even better way, as shown in diagram C on the previous page. If Malfatti had considered a much thinner triangle in the first place, he would have got the answer right first time.

Treacherous triangles

This case illustrates the fact that diagrams can be a source of problems. Drawing a diagram to help with a theorem about all triangles always involves choosing a specific triangle, so there is the danger of finding a proof that applies only to that particular triangle. Though geometrical diagrams are useful in many ways, they can sometimes be more trouble than they are worth. David Hilbert (see more, page 121) was well aware of the pitfalls that diagrams could introduce. In 1899 he managed to develop a version of Euclid's geometry in which no diagrams were needed at all. All the jobs they did were now performed by a set of 21 axioms, with which every theory in Euclid's geometry could be proved.

Human failings

Or maybe not. In 2003, mathematicians attempted to transform Hilbert's geometry into an online version, and discovered that many of his proofs rely on diagrams after all. He had fallen into exactly the same trap as Malfatti. Although Hilbert included the diagrams only to illustrate the proofs, and then went on to spell out all the axioms and other mathematical statements needed to make the proof work, it is sometimes necessary to use the diagram to work out how exactly the statements should be used. Although many hundreds of people had read, studied, and checked Hilbert's work in the century since it was written, it was not until a computer was given the task, and instructed to ignore the diagrams, that the flaws were discovered. Diagrams can be sneaky as well as dangerous.

How it works

Another kind of geometrical error is caused by optical illusions. In this diagram, which of the blue areas looks larger? We can easily work out their sizes, because the 5 rings here are equal distances apart. Imagine them as a series of discs, smaller ones on top of larger. If the small disc in the center has a radius of 1 inch, its area must be π square inches (because the area of a disc $= \pi r^2$). The other discs have radii of 2, 3, 4, and 5 inches, and therefore areas of 4π, 9π, 16π, and 25π. We find the areas of the rings by subtraction: the area of the first ring = area of second circle – area of first circle $= 4\pi - \pi = 3\pi$, and so on. So, the areas of the rings are 3π, 5π, 7π, and 9π. The inner dark area, which is composed of the small circle (area π), and the first two rings, has an area of 9π, the same as the dark outer ring. But is that how it looks?

Below and to the right are more optical illusions where geometrically equal items are distorted.

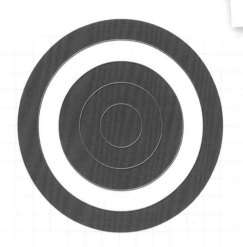

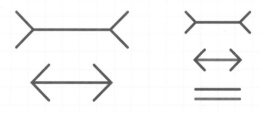

Which horizontal line is longer?

Which blue circle is larger?

SEE ALSO:
▶ Perspective, page 76
▶ The Four-Color Problem, page 140

Non-Euclidean Geometry

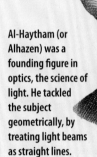

AS THE CENTURIES PASSED AFTER EUCLID'S TIME, many people attempted to prove the parallel postulate in *Elements of Geometry*—namely that two lines could continue forever without meeting. One was the Arab scholar Ibn al-Haytham, who was known as Alhazen in Western Europe.

Al-Haytham (or Alhazen) was a founding figure in optics, the science of light. He tackled the subject geometrically, by treating light beams as straight lines.

Al-Haytham was born in Iraq in about 965 CE, and soon became well known locally thanks to his skill in applying mathematics to practical problems. But it is unwise for even the greatest mathematician to be too confident in their skills, and Al-Haytham's boast that he could control the waters of the mighty Nile River through the power of geometry almost led him to disaster. His claim reached the ears of Caliph al-Hakim, who sent him to Aswan in Egypt to put his ideas into practice. Not surprisingly, he was unable to control the Nile and quickly left for a new job in Cairo instead. When that job, too, proved a failure, the local emir confiscated all

Al-Haytham's possessions as a punishment, and he might have suffered a horrible fate if he had not pretended to go mad—and even then he was kept under house arrest for years, until the emir was safely dead.

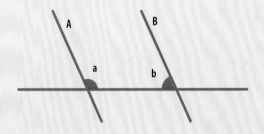

If lines A and B are parallel then the angles a and b must add up to 180°.

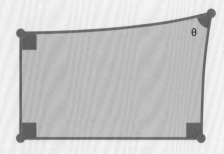

A Lambert quadrilateral is a four-sided shape in which only three of its angles are always right angles.

When he wasn't busy convincing people he was mad, Al-Haytham spent his now very considerable amount of free time doing what any genius would—making scientific breakthroughs. Among other things, he tried to prove Euclid's fifth postulate and seemed to have succeeded.

Strange shape

To do so, he defined a new kind of quadrilateral, in which three of the corners are right-angles but the fourth may be any angle. If you try drawing such a shape, the fourth angle will always turn out to be a right angle too, but Al-Haytham realized that if he could prove this then that would prove the parallel postulate (see pages 48 and 130). However, analysis of his work shows that his proof only worked by assuming the postulate was true, so in fact it led nowhere.

The fourth angle

The same idea was explored by the Swiss mathematician Johann Lambert in 1766. Like Al-Haytham, Lambert wanted to prove that the fourth angle had to be a right angle, because then he could prove the parallel postulate. But first, he explored the implications if the fourth angle was smaller or larger than a

Johann Heinrich Lambert was a seminal figure in hyperbolic geometry, where the sum of the angles in a triangle is less than 180°.

On the surface of a sphere, larger triangles have larger angles. If we lived in an elliptical world, all triangles would behave like this, even those drawn with rulers on paper.

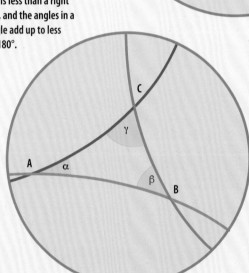

The other new geometry is hyperbolic. In it, the fourth angle is less than a right angle, and the angles in a triangle add up to less than 180°.

he used depended only on their shapes, not their sizes. So, he was able to compare triangles just a few inches long with others the size of the Earth. This would have been impossible in Lambert's strange new geometry.

Bending the rules

To Lambert's great credit, although he hated the idea that the most basic rules of geometry might be wrong, he recognized that his personal feelings did not matter, saying "But all these are arguments dictated by love and hate, which must have no place either in geometry or in science as a whole." While a few other mathematicians became interested in these

Hyperbolic surfaces are concave, meaning they have a negative curvature.

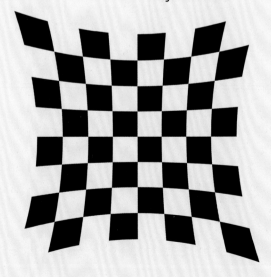

right angle. He found that if it was smaller, then (among other strange things) the sum of the angles in a triangle would be less than 180°. The exact answer would depend on the size of the triangle. But this contradicted a very basic assumption. For instance, when Eratosthenes measured the Earth (see more, page 50), he did it by assuming the angles in the triangles

Right: The Mad Hatter's tea party in *Alice's Adventures in Wonderland*. Its impossible chaos was partly an attempt by the author Lewis Carroll (who was also a mathematician) to illustrate what he regarded as the silliness of some new ideas in math, including non-Euclidean geometry.

ideas, it was not until 1830 that this new geometry was studied properly. After all that time, two separate mathematical studies turned up at once, one by the Hungarian mathematician János Bolyai and the other by Nikolai Ivanovich Lobachevsky, a Russian.

Two new worlds

Because Euclid's geometry is largely based on the parallel postulate, the geometries of Bolyai and Lobachevsky are called non-Euclidean. There are two such geometries, hyperbolic and elliptic. In hyperbolic geometry, the fourth angle in a Lambert equilateral is less than 90°. In elliptic geometry, the same angle is greater than a right angle, and the angles in a triangle add up to more than 180°. Shapes on the two-dimensional surface of a sphere actually obey these rules, so in fact map-makers had been using the principles of elliptic geometry for centuries. However, hardly anyone had considered that the world around them could be like this, since it would lead to oddities such as drawing a triangle on a flat piece of paper, and finding that its angles were more than 180°. In fact, there would be no such thing as flatness, nor straight lines. The world would

not be as it appears because light rays would not be straight, so shapes would shift as you approached them.

Defects

Any mathematician, faced with a strange new world, will search for a way to quantify it—that is, to pin it down with numbers. One way to do this is by defining the "defect" of a triangle. The defect is the difference between 180° and the angles in a triangle, and it is negative in a hyperbolic world, zero in a Euclidean one, and positive in an elliptic world. The maximum possible deficit for a triangle is 180°. In this case, its angles are zero and its axes are infinite—yet it has a finite area. Lewis Carroll (who wrote *Alice in Wonderland*) was a keen mathematician, but this fact about triangles made him certain that non-Euclidean geometry must be nonsense.

Although the embarrassing error of Malfatti (see page 126) shows how dangerous diagrams can be, Lewis Carroll's disbelief about triangles shows how useful they are, because we can easily sketch them. You just have to bear in mind that non-Euclidean triangles would always have sides that look absolutely straight. It is the spaces they occupy that creates the curvature.

Meanwhile, in the real world

You may well be wondering, as many mathematicians did wonder in the 1830s, what the point of all this is, given that we live in a Euclidean world? Well, perhaps we do not. Bearing in mind that in a non-Euclidean world, rulers and lines and light rays all seem perfectly straight, and that no one has actually checked whether the angles of gigantic triangles are really the same as those of teeny ones, so how can we be sure of this? We would soon notice if the defects of triangles were large, but what if they are very small? Bolyai's conclusion was that the question was unanswered. It was the job of physicists, not mathematicians, to discover what the real world was like. Nearly a century later, one of them did.

How it works

Absolutely correct

It is possible to construct a kind of geometry that makes no use of the parallel postulate at all, which means it is not hyperbolic, Euclidean, or elliptical. This is called absolute geometry, and, in fact, some of Euclid's proofs are absolute. They are true whatever space you try them in. One of them is, if you subtract one of the angles in a triangle from 180°, the angle that results (called an exterior angle) must be greater than either of the other two internal ones. Sadly, although this is actually true, Euclid's proof is one of the few that he got wrong.

In any triangle in any space, angle A is larger than angle C and also larger than angle D.

Hyperbolic parallel lines

Parallel lines look different in the three geometries.

Euclidean parallel lines

Elliptic parallel lines

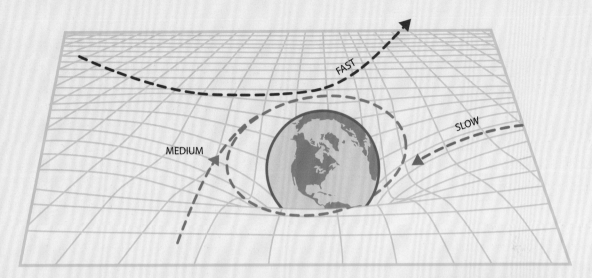

The Earth's mass curves flat space around it. This space warp creates the effects of gravity that we observe on moving bodies.

Strange but true

In 1919, Albert Einstein formulated a theory which explained gravity as a curvature of space and time. To understand what this means, imagine that our Universe only has two dimensions, like a sheet. The sheet is flexible, as rubber is, so any objects placed on it make it sag a little. A very massive object such as the Earth causes a major sag. Imagine what happens if a space capsule is flying close by the Earth (above). Because the rubber it is crossing is curved downward, the capsule will tend to follow that curve. It may fall right down to the Earth, or if it is going fast enough, it might fly straight past, but be deflected off course. Or, if the speed is in between these two, it might go into orbit. We can explore this space by drawing a Lambert quadrilateral on it (right). The angle θ in the lower left corner of the quadrilateral is less than 90°, which means that space is hyperbolic.

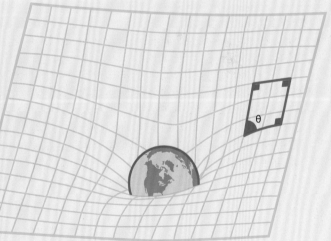

The mass of Earth has made space around it hyperbolic.

SEE ALSO:
▶ Euclid's Revolution, page 44
▶ Flat Maps of a Round World, page 82

Crystals

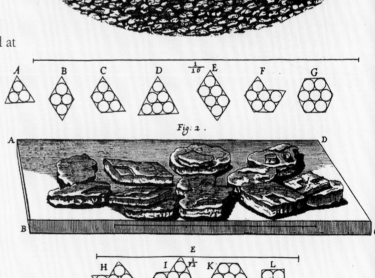

Fig: 1

IN 1665, ROBERT HOOKE PUBLISHED ONE OF THE MOST BEAUTIFUL SCIENCE BOOKS EVER WRITTEN, CALLED *MICROGRAPHIA*. It recorded in woodcuts what Hooke had seen though the microscopes he had made. This was literally a new world, never seen before, and people were amazed at what everyday creatures such as fleas and flies really looked like.

Among many other things, Hooke looked at crystals, and he was struck by the fact that they looked even more neat and geometrical at high magnifications than they did when seen with the naked eye. He wondered, as Harriot and Kepler had wondered before him (see more, page 92), whether they might have these shapes because they were made of stacked-up spherical "corpuscles" (we might use the word atoms today), just as a pyramidal shape can be made by stacking up cannonballs.

Haüy's lucky break

In the 17th and 18th centuries, people were fascinated by the strange properties of calcite, a calcium carbonate mineral that is often

Hooke's *Micrographia* included close-up drawings of crystals, as well as the author's proposals for how they were constructed at the atomic level to create their repeating features.

transparent. Objects were seen to double when viewed through thin slabs of calcite, and the distance between the double images depended on how the crystal was positioned. In 1781, René-Just Haüy, a young student of natural history, was examining an especially beautiful piece of calcite when he dropped it. It shattered into many pieces, and he was fascinated to see that all the fragments had similar shapes, all based on rhomboids. Haüy experimented with other minerals and found that many had their own characteristic crystal forms, which could be revealed by tapping them just hard enough to break them. He also found ways to stack cubes to form some of the crystal shapes he found. It was soon clear that there were many different crystal forms. How should they be classified? The classification of shapes is the kind of challenge

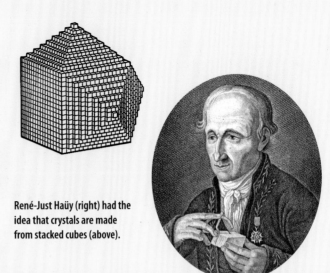

René-Just Haüy (right) had the idea that crystals are made from stacked cubes (above).

that geometers love, so many of them set to work on the problem. The basic idea is the same as that used to classify tiling patterns, which is to group crystals according to the kinds of symmetry they have, as shown below.

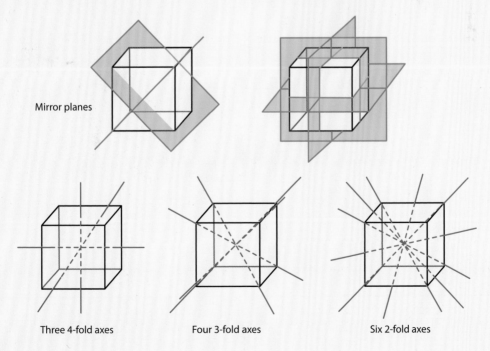

Mirror planes

Three 4-fold axes   Four 3-fold axes   Six 2-fold axes

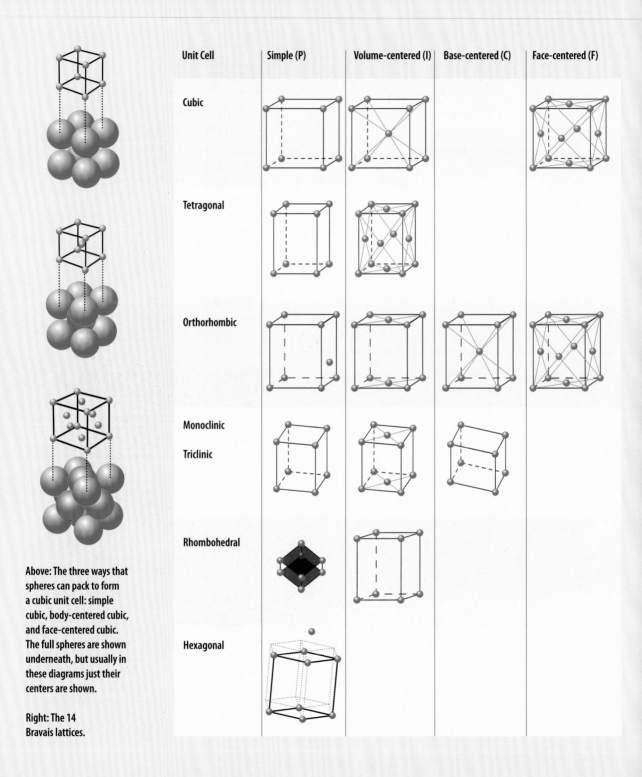

Unit Cell	Simple (P)	Volume-centered (I)	Base-centered (C)	Face-centered (F)
Cubic				
Tetragonal				
Orthorhombic				
Monoclinic Triclinic				
Rhombohedral				
Hexagonal				

Above: The three ways that spheres can pack to form a cubic unit cell: simple cubic, body-centered cubic, and face-centered cubic. The full spheres are shown underneath, but usually in these diagrams just their centers are shown.

Right: The 14 Bravais lattices.

Bravais Lattices

There were soon so many ideas proposed to make sense of crystal structures that today we have a quite complicated system that gives a very complete picture. The best early system was developed by Auguste Bravais, a young French mathematician who was so clever that in 1829 (aged just 18) he won first prize in mathematics in a competition, which helped him attend a famous college in Paris called the École Polytechnique (where many other mathematicians also studied). He was soon top of his class, which meant he was free to choose any technical field he liked. He chose to join the navy. Though he loved math, he loved the idea of traveling the world even more, and he spent the next few years mapping Africa and then exploring the Arctic. During the long, dark winters in the far north, Bravais worked on the classification of crystals. By 1848, he had defined seven "unit cells." A unit cell is the smallest part of a crystal to show its system. The simplest of these, and the most symmetrical, is the cube. Each of the seven systems can be broken down further, because there is more than one way to stack spheres to produce a unit cell. In total, Bravais found that there are 14 ways to stack spheres. These ways are called Bravais lattices.

SEE ALSO:
▶ Circles and Spheres, page 14
▶ Tiling and Tessellations, page 70
▶ Filling Space, page 92

POLYMORPHS

Two (or more) crystals are polymorphs if they are made of the same chemicals, in the same proportions, but use different Bravais lattices. Knowing the difference can be important because polymorphs may have very different properties. In 1912, the explorer Robert Scott and his whole team died on their way back from the South Pole. This was partly because their tins of heater fuel had mysteriously emptied. The tins were sealed with lead, which, at low temperatures, shifts from its strong metallic form to a weaker polymorph which let the oil leak out. Carbon has several polymorphs, as seen here, which include graphite, the main ingredient of pencil "lead," soot, and diamonds.

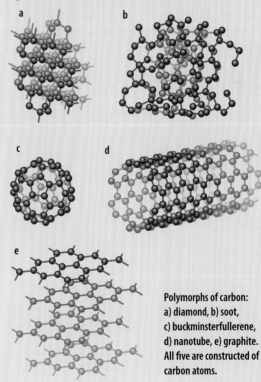

Polymorphs of carbon:
a) diamond, b) soot,
c) buckminsterfullerene,
d) nanotube, e) graphite.
All five are constructed of
carbon atoms.

The Four-Color Problem

A map of the counties of England and Wales from 1862 uses four colors.

IN 1852, A YOUNG MATH STUDENT CALLED FRANCIS GUTHRIE WAS PONDERING A MAP OF ENGLAND. He was wondering how few colors were needed to color the different counties, so that no neighboring counties have the same color. After plenty of trial and error, Guthrie decided that four would be enough. All he had to do was to prove it, but he didn't know enough geometry to do so.

Guthrie asked his teacher, Augustus De Morgan, for help with this problem. De Morgan was a highly skilled mathematician, but he couldn't manage it either, and neither could any of his colleagues. By 1878, the problem had become famous enough to appear in the science journal *Nature*. Since it was such a simple problem to understand, surely it must be fairly simple to solve?

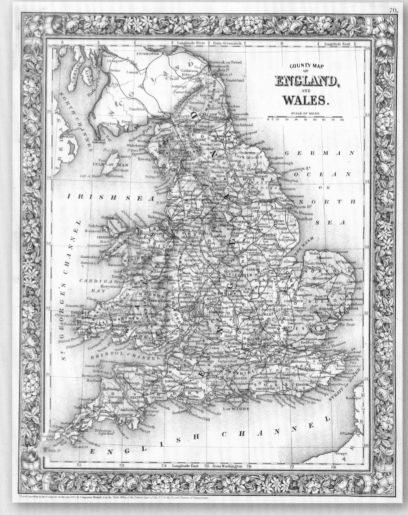

Possible proof

But it wasn't only Guthrie who didn't know enough geometry—no one did. A year later a mathematician called Alfred Kempe seemed to have proved the four-color conjecture, as it had become known, but 11 years after that a flaw was found in Kempe's proof by a mathematics lecturer at Durham University, Percy Haewood.

Five is enough

Haewood was famous locally for the age of his cape, the size of his mustache, and the charm of his dog (who usually came with him to his lectures), and he made the first real step forward in tackling the problem by proving that five colors would definitely be enough. However, no one could find an example of a map that actually needed five colors. To do this meant studying the pattern of every possible map there could be, which would involve checking millions of complex graphs. How could anyone do this? And even if it was done, how could anyone be certain that not a single mistake had been made?

Math by machine

Finally, in 1976, the theorem was proved by mathematician and computer expert Kenneth Appel and topologist Wolfgang Haken. A computer was needed to carry out the ten billion calculations involved, and it took about

The computer that proved the theorem, an IBM 370-168.

GRAPH THEORY

To geometers, a graph is a web-like pattern used to simplify maps, mazes, and other complicated diagrams. Graphs allow us to focus just on the things that matter to us about the map, which in this case is the boundaries between countries. We can ignore things that don't matter, like shapes and sizes.

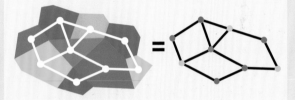

A map and its corresponding graph.

1,200 hours of processing time. It was the world's first proof by computer, and many mathematicians were very unhappy about that, because no human can ever check the proof and see that is right. Could computers replace mathematicians? Not yet, for although today's computers are many times more powerful than the ones that proved the four-color theorem, they very rarely make decisive contributions to proving new theorems. So mathematicians are safe—at least for the moment.

SEE ALSO:
▶ Knots, page 146
▶ Fractals, pages 166

The Roundness of Rollers

THE WHEEL IS OFTEN SAID TO BE THE MOST SIGNIFICANT INVENTION IN HUMAN HISTORY. The roundness of the wheel makes it roll along and move loads that exceed the abilities of muscle alone. However, does a wheel need to be round?

It is assumed that rollers of some kind were used to build ancient edifices such as Stonehenge (see page 16). While the precise technology used is open to guesswork, geometry can tell the limits of how round a roller needs to be.

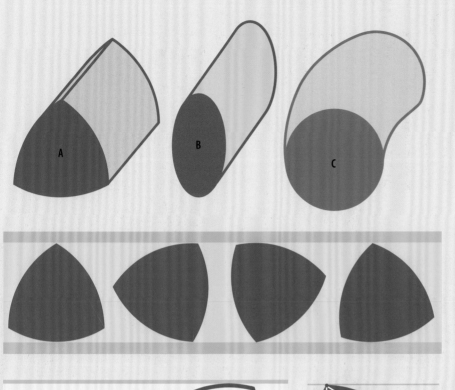

Rollers need not always be perfect cylinders. Which of these rollers would do the job best?

We can discount Rollers B and C. They will give a rough ride, but roller A is able to turn without any up or down movements.

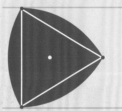

Far left: The center of the roller moves up and down, while the height does not.

Left: The roller is a triangle with curved sides.

Straight and level

Before there were wheels, which spin around a central rod called an axle, there were rollers, which are not fixed to anything. If you had to use a set of rollers shaped like either A, B, or C, shown above, which would you choose to make sure your heavy slab of stone stayed level as you rolled it along? Roller A looks the least round so perhaps not that one. However, Roller B has an oval shape so it will rise up and down, and Roller C may have a round cross-section but is curved along its length so will not roll at all. Perhaps we should look at A again. Counter-intuitively, a set of A Rollers would give a perfectly level ride (if the ground were flat). The only reason that we don't use shapes like this as wheels on cars and bikes, for example, is because the center of the shape does shift as it rolls, so any axle would move up and down as the vehicle moved, creating a very bumpy ride.

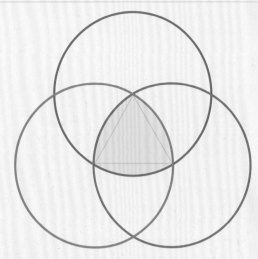

The Reuleaux triangle, above, named after Franz Reuleaux, right, is formed from overlapping circles.

Reuleaux triangle

There are many shapes that behave like this, and they have many uses in engineering. In fact, the particular shape above is named after a German engineer, Franz Reuleaux, who used it in many different designs in the 1870s. A Reuleaux triangle can be defined as the edges of the overlapping areas of three circles. The secret of the triangle is that it has the same width, no matter how you measure it, and because of this it is known as a "curve of constant width."

Geometry in action

Machines which accept coin payments work out which coins have been added by measuring their

A Canadian "Loonie" dollar and a British 50p coin are both curves of constant width.

There are other shapes with similar properties to the Reuleaux triangle. While the Reuleaux triangle spins exactly within a circle, the shape below, called a delta-biangle, rotates inside an equilateral triangle. (See more, box opposite).

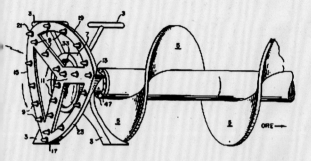

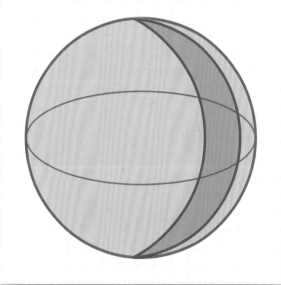

How it works

Biangles

The biangle is a shape which does not exist in Euclidean geometry, but is one of the simplest shapes in spherical geometry, which is the geometry of shapes on the surface of a sphere. The angles at the ends of a biangle are equal and the lengths of its sides are equal too. If the angles of a biangle are measured in radians, then the area of a biangle equals twice its angle. (There are 2π radians in a circle, so $360° = 2\pi$ radians, so one radian = $(360°)/2\pi \approx 57.3°$; see more, page 23).

A Reuleaux triangle drill bit can dig out a square hole.

widths (as well as by weighing). So, any non-round coins have to be curves of constant width. A Reuleaux triangle can rotate happily in a square hole, with the center point moving in a loop. That means that a Reuleaux-shaped drill bit can be used to drill square holes, as long as its center is allowed to move in a small circle as the rest of the drill bit rotates.

SEE ALSO:
▶ Archimedes Applies Geometry, page 52
▶ Flat Maps of a Round World, page 82

Knots

KNOTS HAVE BEEN USED FOR PRACTICAL AND DECORATIVE PURPOSES SINCE LONG BEFORE WRITTEN HISTORY BEGAN. The oldest fragment of rope is around 28,000 years old. However, mathematicians only began to study them in terms of their geometry in the 1870s. Even then, it was only because knots seemed so useful for exploring the latest atomic theory.

It was physicist William Thomson, often better known as Lord Kelvin, who developed the new theory of matter, based on the idea that atoms were really knots of a material called aether (see more, page 34), which was thought to fill all of space. Thomson turned to his colleague and friend, mathematician Peter Guthrie Tait, for help in understanding knot mathematics, but Tait soon realized that hardly anyone had studied the subject before. In 1878, Thomson had attended a lecture given by Tait in which Tait demonstrated

Some of the many knots that have been developed for different purposes over the centuries.

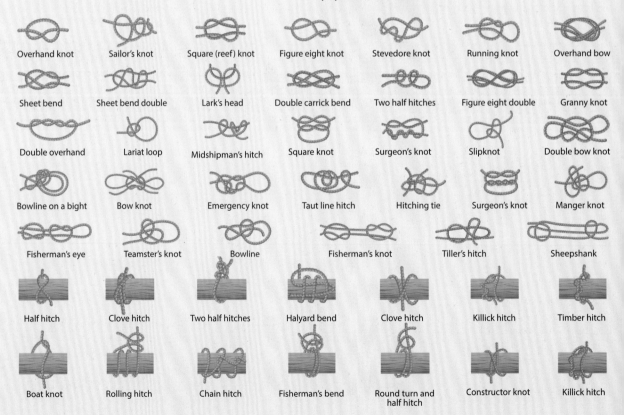

Overhand knot	Sailor's knot	Square (reef) knot	Figure eight knot	Stevedore knot	Running knot	Overhand bow
Sheet bend	Sheet bend double	Lark's head	Double carrick bend	Two half hitches	Figure eight double	Granny knot
Double overhand	Lariat loop	Midshipman's hitch	Square knot	Surgeon's knot	Slipknot	Double bow knot
Bowline on a bight	Bow knot	Emergency knot	Taut line hitch	Hitching tie	Surgeon's knot	Manger knot
Fisherman's eye	Teamster's knot	Bowline	Fisherman's knot	Tiller's hitch	Sheepshank
Half hitch	Clove hitch	Two half hitches	Halyard bend	Clove hitch	Killick hitch	Timber hitch
Boat knot	Rolling hitch	Chain hitch	Fisherman's bend	Round turn and half hitch	Constructor knot	Killick hitch

the knot-like behaviors of smoke rings. It was this that gave Thomson the idea that atoms might be knotted loops of aether.

Knots and unknots

With the job of developing a whole new area of mathematics on his shoulders, Tait had to start at the beginning by listing all known knots. Like other knot mathematicians since, Tait found it simplest to consider knots as closed loops, and to include in his list an unknotted loop called an "unknot." He then listed all other knots, starting with the simplest, the trefoil. The most obvious way to classify the trefoil is to count how many times the string crosses itself, which is three. The first difficulty with this approach became obvious as soon as Tait reached a knot with five crossings. There are two different knots of this kind, and after that the number of different knots with the same number of crossings increases rapidly. So the simple crossing count offers no protection against overlooking knots. So far, geometers have classified all knots with 16 or fewer crossings. That may not sound very impressive, but that already includes 1,702,936 knots. However, the big problem with classifying knots by counting their crossings is that there are many ways to twist a knot to add lots of temporary crossings.

That which remains the same

In 1885, the American mathematician and engineer Charles Little listed what he thought

Above: A trefoil knot, depicted here in an infinite loop, is named after a flower with three petals.

Right: The first 15 members of Tait's knot table, including the unknot, or loop. The main number shows how often the string crosses itself, and the subscript says which version the knot is when there are more than than one with the same number of crosses.

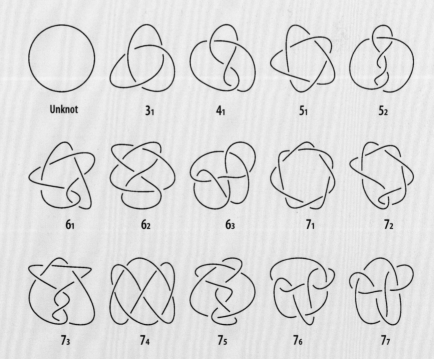

Unknot 3_1 4_1 5_1 5_2

6_1 6_2 6_3 7_1 7_2

7_3 7_4 7_5 7_6 7_7

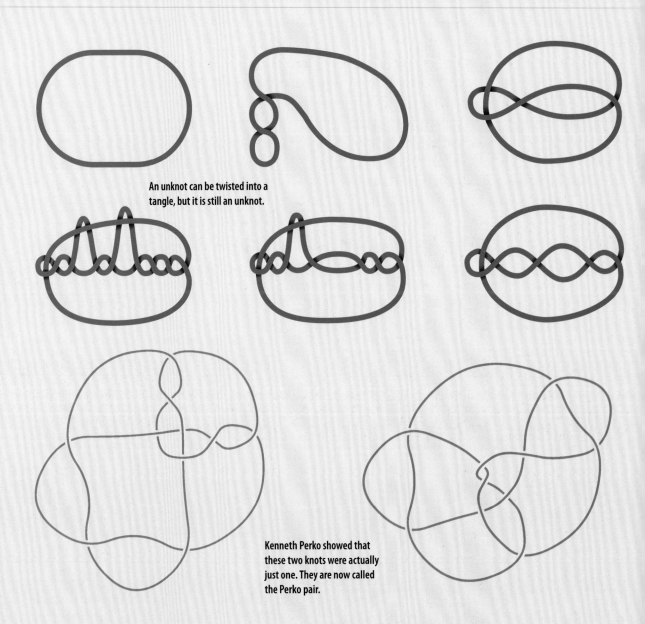

An unknot can be twisted into a tangle, but it is still an unknot.

Kenneth Perko showed that these two knots were actually just one. They are now called the Perko pair.

were all the knots with up to 10 crossings. There were 166 of them. Nearly a century later, in 1974, Kenneth Perko, an amateur mathematician who spent most of his time being a New York lawyer, realized that two of Little's knots were actually the same. Even though they were right next to each other in Little's table, no one had spotted this before. Actually, Perko didn't spot them either: eyes are of little use in analyzing knots, and one of the most brilliant knot mathematicians, Louis Antoine, was blind. Antoine developed a kind of infinite knot now called Antoine's necklace (see below and the box on page 150). Perko only came across the double-counted knot when he was checking whether Little's method of knot analysis worked. The fact that the method classified the same knot as two different ones showed that it didn't!

The search for invariants

What Charles Little had tried and failed to do was to find an invariant. A knot invariant is a number, equation, or measurement which can be applied to any image of any knot. No matter how twisted a knot is, its invariant must remain

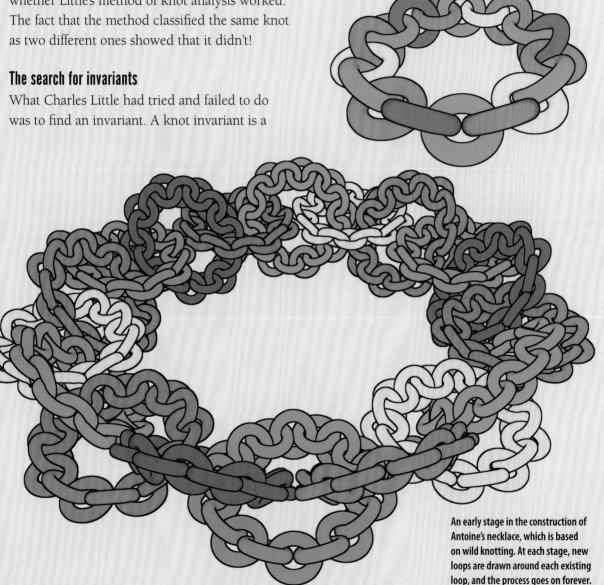

An early stage in the construction of Antoine's necklace, which is based on wild knotting. At each stage, new loops are drawn around each existing loop, and the process goes on forever.

A WILD KNOT

One way to test any mathematical theory of knots is to ensure that the theory does not lead to wild knots, which contain infinite loops.

The simplest wild knot.

the same, and no matter how similar a pair of different knots look, their invariants must be different. For a while, it seemed that the problem of knot types had finally been solved in 1970, by Wolfgang Haken. Haken came up with a new way of looking at the knot analysis problem (and looking for a new approach is always a good idea when you work on an old problem). We can get an idea of Haken's approach by dipping some loosely knotted loops of wire into bubble liquid. Each will emerge from the liquid with a soap film stretched across it, and Haken saw that mathematical versions of surfaces a bit like these should work perfectly as a means to describe unique knots. Haken was a programming genius and he was working on a computer program to run his knot test when he stopped to work on the four-color map problem (see more, page 141). His program needed further work and was only completed in 2003. It may seem strange that knot theorists hadn't rushed excitedly to get it working long before this, but a new problem had become obvious almost as soon as Haken

began to develop the program. Running his knot test would take so long that no computer in the world could do it in a reasonable time. Despite the increases in computer power since 1970, the same is true today. So, the search for a foolproof and usable tool for analyzing knots continues.

Down to Earth

Although the knots that mathematicians work with don't seem of much relevance to everyday ones, one very everyday question has been answered by knot theory: is a bundle of strings really more likely than not to get tangled into a knot, or is it just bad luck when it does? The answer came in 1988, from a mathematician called De Witt Lee Sumners and a chemist, Stuart Whittington, who were interested in the way knots form in long, string-like molecules called polymers. They worked by carrying out a mathematical version of this exercise: imagine you are playing a real life game of snakes and ladders in a multi-story building. You throw a die for each move. If you throw 1, 2, 3, or

4 you must move North, East, South, or West respectively. If you throw a 5 you climb up the nearest ladder to the next floor, and a 6 means you slide down the nearest snake. You track your progress by uncoiling a rope as you go, and the only rule is that you cannot pass the same point twice. The game is finished either when you have been to every point in the building, or you are surrounded by spaces you have already visited. In either case, Sumners and Whittington found that you are much more likely than not to have made at least one knot in your rope by the time you finish. So, bundles that get knotted together are more or less inevitable.

SEE ALSO:
▸ Perspective, page 76
▸ Topology, page 122

CHEMICAL KNOTS

The idea that atoms might be knots of matter did not last long, and for many decades after that, only a few mathematicians were interested in the geometry of knots. But in the last few years, knots have become of great interest to scientists once more. In nearly every cell of your body are strands of a molecule called DNA (deoxyribonucleic acid), which contain all the instructions your body needs to live and grow. But, while a cell is too small to see, its strand of DNA is about 5 feet long. To fit, it must be bundled up tightly, and so it is full of knots. To use the DNA to make new cells, sections of it must be cut, unknotted, and spliced together again. Complex chemicals called enzymes do this, and understanding how they work should help doctors tackle some of the many diseases that are caused by damaged DNA. But, no one has been able to see this process taking place. All we can do is to study the DNA's knotted shape before and after the enzymes have worked on it. However, if geometry could define the before and after in terms of knot types, it should then be possible to work out what the enzyme has done in the intervening period.

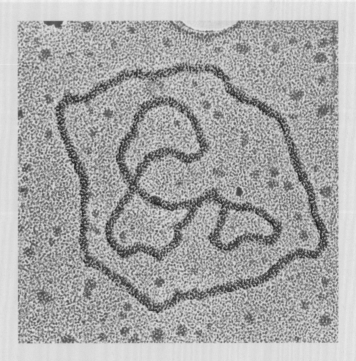

An electron microscope image of DNA. Despite their initial appearance, these long molecules have a highly coiled structure.

The Poincaré Conjecture

MOST MATHEMATICIANS MAKE NEW BREAKTHROUGHS BECAUSE THEY LOVE MATHEMATICS, WHICH IS JUST AS WELL SINCE THERE ARE FEW FINANCIAL REWARDS FOR NEW THEOREMS. But in 2000, a mathematical organization called the Clay Institute decided to do something different. The institute chose seven of the most important unsolved mysteries of mathematics and offered a $1 million prize to anyone who could solve any of them. Only one of these seven "Millennium Problems" has so far been solved, by a Russian mathematician called Grisha Perelman. But he refused the cash prize ...

Henri Poincaré left us a big question about shapes and space.

The problem is called the Poincaré conjecture, named after its inventor, Henri Poincaré. He was one of the greatest mathematicians of his time. In a school test he was asked to sketch a pair of curves of different kinds and talk about how they might overlap if viewed from the correct angle. Poincaré solved an extended version of the problem, using mathematics and not trial and error. However, he drew his answer upside down, so although it was perfect, it was marked as wrong. As an adult, too, Poincaré was often distracted by concentrating on mathematical problems. For one of his many foreign trips he packed his bedsheets instead of his shirts.

Kinds of space

Poincaré's conjecture concerns the topologies of different kinds of space. Topology can be a confusing subject because it uses a lot of unfamiliar words, and some familiar ones in unusual ways (see box, right). In everyday language, the Poincare conjecture says that every three-dimensional surface with no holes can be distorted into a three-dimensional spherical surface. It is easy to check that a circle (which is a one-dimensional line) has no breaks or holes in it. We can simply glance at one, laid out in two-dimensional space, such as on a piece of paper.

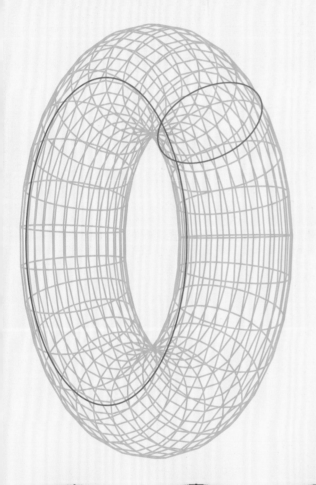

SURFACES

In ordinary language, a surface is two-dimensional, like the flat surfaces of the walls and floor of a room, or the curved surface of a ball. But in topology, lines are often called one-dimensional surfaces, and what we would usually call volumes, or spaces, or interiors, are called three-dimensional surfaces. This is because all these different things (and their higher-dimensional versions, too) can have similar characteristics. One of these characteristics is that a surface can either be open or closed. A circle, or the skin of an orange, or the inside of a football, are examples of closed surfaces in different dimensions. On a graph, the x and y axes go on for ever, so they are both open one-dimensional surfaces. The space between those axes, or an endless, flat plane, would be open two-dimensional surfaces. The other characteristic that defines a surface is its curvature. An ellipse is a curved one-dimensional surface, an eggshell—although it surrounds a three-dimensional shape—is a curved two-dimensional surface.

The Poincaré conjecture asks how to count the holes in a geometric figure. For shapes in one, two, or three dimensions, like a donut or sphere, the answer is easy to check. But what happens in higher dimensions?

Poincaré tried to answer his question by imagining string looping around objects. If the loop closed to a point on the surface, there were no holes in the object. Again, that seems easy. Now try it again in four dimensions or more!

 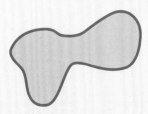  

How many shapes do you see? Poincaré and his fellow topologists just see one—a circle.

It is also easy to find out that a spherical surface (which is two-dimensional) has no holes in it. Like us, it exists in three-dimensional space, so we just need to look at it from all angles. If we look at the surface of a donut, it is just as easy to see that there is a hole it in, as long as we look at it from all angles. However what if we go up another spatial dimension? Then we are faced with a three-dimensional surface. If we could somehow look at it from different angles in four-dimensional space it would be obvious whether it has a hole in it or not. But we can't see it this way.

A view from the fourth dimension

To overcome this problem, in 1904 Poincaré came up with a hole-counting method called homology. To see how it works, imagine a two-dimensional surface, like the surface of a ball. If we take a piece of string, we can loop it around the ball in any way we like, and it will always be possible to tighten that loop to a point or knot on the surface without keeping any of the object within the loop. But on a donut we can arrange the loop through the hole so it can't be tightened to a point or knot without cutting through the donut. Poincaré suspected that this test would work on any three-dimensional surface, but he couldn't prove it.

Space traps

One of the challenges is that we want a test that works even for distorted shapes. All the lines shown above, for instance, are just different versions of the same one. This is easy to imagine if we think about a loop of string that we could change to look like any of them, without cutting

The hourglass and the bell are the same shape—a kind of sphere—and Poincaré's homology method shows us that.

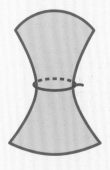   

the string. In a similar way, cubes and pyramids and ellipsoids are all distorted versions of a sphere: we could make any of them from a spherical lump of modeling clay. And we can prove that with our loop: we can always shrink it to a point, whatever the shape of the lump, although it may need a little remodeling first. If we start off with an hourglass shape (see opposite, below) with the loop around the narrow neck, the loop will be trapped, even though the hourglass has no holes in it. This is a space trap. We can escape it by remodeling the shape so we can continue to shrink the loop.

Surfaces beyond view

So, now we know that with two-dimensional shapes like the surfaces of spheres or of hourglasses, we can use Poincaré's shrinking-loop test, but we might need to re-mold the shapes first. But how can we do this with the unimaginable three-dimensional surfaces of four-dimensional shapes? If our loop can't contract, how can we tell whether that is because it has fallen into a space trap or because the surface has a hole in it? And, even if we know that a space trap is our problem, how can we re-mold our four-dimensional shape to free our loop, given that we can't visualize what the shape looks like?

Richard Hamilton

Smoothing

The solution was found by an American topologist called Richard Hamilton, who developed a method of re-molding four-dimensional shapes automatically, without the need to visualize them. He did this by treating curvature like heat! Heat has an unusual property in that it evens itself out. If we hard-boil an egg, we might put it into cool water afterward. To begin with, most of the heat is in the egg, but after a while, the heat has evened itself out, so instead of a hot egg and cool water, we have a warm egg and warm water. What has that got to do with shapes? Let's start with the three below.

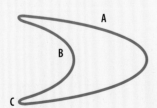

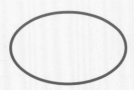

These three shapes are the same to a topologist, but their curvatures are not evenly spread. This is the root of the space trap problem.

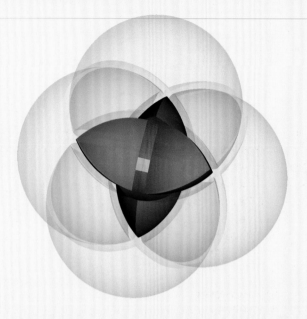

The red space is formed by the space shared by four intersecting spheres. This is one way of visualizing the three-dimensional surface of a four-dimensional cube or tesseract.

We can see that the first is almost flat at A, slightly curved at B, and very curved at C. So, the amount of curvature it has is very unevenly spread. The football shape in the middle is more evenly curved, but still the ends have more curvature than the middle. Only in the sphere at the end is the curvature spread evenly. If Hamilton could find the secret of making curvature behave like heat, then he could make shapes smooth themselves out just as heat automatically evens itself out. In the 1980s, Hamilton found just the right mathematical tool for the job, a process with the odd name of "Ricci flow" (see box opposite). The method was named by Hamilton after the Italian mathematician Gregorio Ricci-Curbastro who had made important contributions to the geometry of curved spaces in the 1880s. Hamilton was well on the way to using his Ricci flow technique to solve the Poincaré problem, when he encountered a new kind of trap altogether.

The challenge of infinity

This trap involves the way certain curves head for infinity and become unsmoothable. If you investigate $y = 1 \div x$ by starting at $x = 5$, counting down by 1s, and joining the dots, a smooth and simple curve appears (see overleaf). Until, that is, you reach $x = 0$. $1 \div 0$ is infinitely large, so trying to plot it on the graph gives an endless upward spike. This is called a singularity, and if one appears on a surface it will wreck any flow

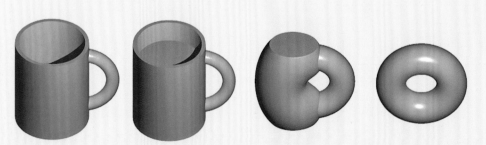

Breakfast for a topologist is confusing because they see no difference between the coffee mug and a donut.

How it works

Flows

The kind of flow Hamilton used is called Ricci flow and is highly complex. A simpler kind is called curve shortening (or reducing) flow. The process begins by measuring the curvature of the shape to be smoothed. This is done by fitting circles as closely as possible to the curves in the shape. The tighter the curve is, the smaller the circle needed to fit it.

Next, a speed is worked out for every point where a circle touches the shape. A big circle gets a low speed, a small one gets a faster speed. The blue arrows show the speeds—faster speeds have longer arrows—and each arrow points toward (and perhaps through) the center of its circle. Finally, each part of the shape moves as the arrows indicate, which quickly evens out the curvature.

Smoothing the green shape using curve shortening.

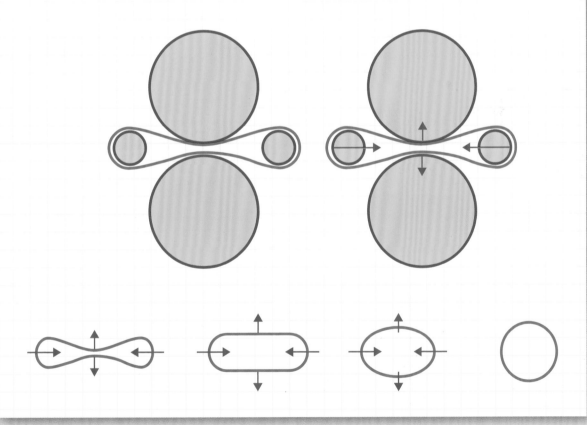

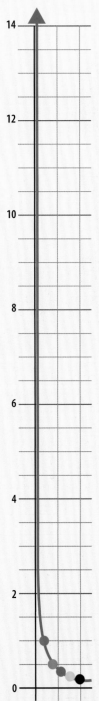

process smoothing things out. The person that solved this problem was a Russian genius called Grigori Perelman. Aged 16, Perelman had received a gold medal for getting every single question right in an international mathematical competition, and had gone on to work on mathematical problems in the United States and back home in Russia. Perelman dealt with singularities just as we might solve our 1 ÷ x problem: if we cut out a region in the middle section of the graph, we get rid of the singularity.

Solved at last

This cut allows us to continue to explore 1 ÷ x on the other side, to the left of the y axis when x turns negative. Like so much else about topology, this technique is simple when you can see what you are doing, but in tackling the world of three-dimensional surfaces, which only a four-dimensional creature could see, it is extremely tough. Among many other problems is the need to be certain that the cut-away areas do not contain the very holes you are looking for.

Below: The line y = 1/x is a simple curve when dealing with the counting numbers.
Left: As the line approaches the y axis (where x = 0), it also approaches infinity.

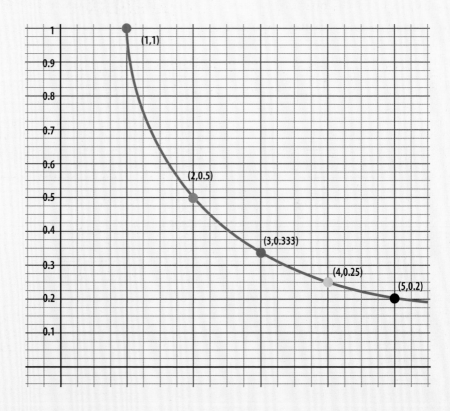

Grigori Perelman is perhaps the greatest mathematician alive today. He chooses to live quietly, away from the spotlight.

However, Perelman managed to do it, and in so doing he proved Poincaré's conjecture at last. So, why did he refuse Clay's $1 million Millennium Prize? Many explanations have been given. It's claimed by some that he said, somewhat strangely, "Emptiness is everywhere and it can be calculated, which gives us a great opportunity. I know how to control the universe. So tell me, why should I run for a million?" But in his official response to the offer of the prize his answer was much simpler and more modest: it would be unfair of him to accept the prize alone, because Richard Hamilton deserved it, too.

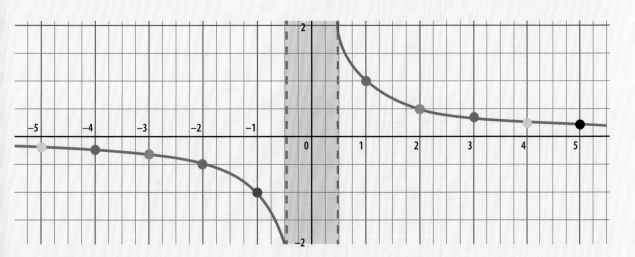

Above: Cutting out the region of the graph where the curve becomes ill-behaved solves the problem of smoothing shapes.

SEE ALSO:
▶ Higher Dimensions, page 116
▶ Topology, page 122

The Geometry of Space and Time

IN 1905, ALBERT EINSTEIN MADE A DISCOVERY THAT CHANGED OUR UNDERSTANDING OF THE UNIVERSE FOREVER. He showed that both space and time change when an object travels fast.

Special relativity, which is the name now given to this discovery, says that if you are on a spacecraft going past Earth at, say, 500 million miles an hour, you will see that the planet and everything on it has narrowed. The rounded globe will be a prolate spheroid. This phenomenon is called relativity because the changes are relative to (that is, they depend on) the observer. Everyone down on Earth (with the right telescope) would see your rocket get shorter the faster it went. But it would seem the same as ever to you. Your

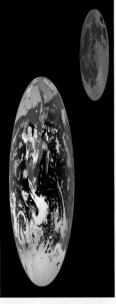

Left: The view from a spacecraft firstly at rest above Earth and then passing by at a considerable fraction of the speed of light. The faster you go, the narrower the outside space will all become.

Right: Similarly, the speeding spacecraft appears to be compressed in the direction of travel when viewed from the surface of Earth.

ruler measures the lengths of objects aboard to be exactly the same as before launch. However, observers on Earth see that your ruler has been compressed along with everything else.

Time as well as space

Something similar happens to time. If there were a powerful light beacon on the North Pole that let out a bright flash once a second, then that rate of flashing would become slower the faster you fly past. If you looked down at the people on Earth, you would see they are moving around in slow motion. Again, this is a relative change. If your

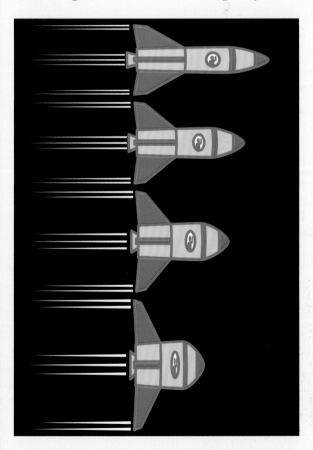

LENGTH AND DURATION

Einstein's relativity theory includes a formula which allows us to work out just how the length of an object changes depending on its speed relative to an observer. $Length_v$ is the length as measured by an observer whom the object moves past at velocity v, $Length_0$ is the length according to someone on board (sometimes called "proper length"), and c is the speed of light.

$$Length_v = Length_0 \times \sqrt{(1 - (v^2 \div c^2))}$$

The equation for the stretching of time is similar

$$Time_v = Time_0 \times 1/\sqrt{(1 - (v^2 \div c^2))}$$

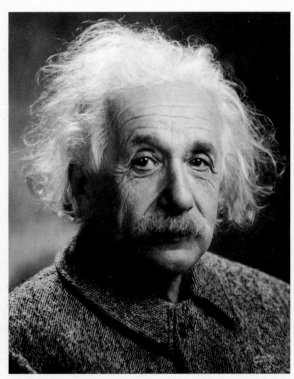

LOW SPEED RELATIVITY

The speed of light is 670,616,629 miles per hour (299,792,458 meters per second), and it is this high speed, together with the squared terms in the formulas, that explain why we don't usually notice the effects of relativity. The fastest human beings were the crew of the Apollo 10 spacecraft, who traveled at 24,791 miles per hour (11,083 meters per second) in 1969. This is less than 0.004 percent of the speed of light, which means that they and their capsule shortened by about 0.00000007 percent, and their watches and hearts and thoughts slowed down by about this same tiny amount.

The world's fastest men, Eugene Cernan, John Young, and Thomas Stafford. Behind them is the Saturn V rocket carrying the Apollo 10 capsule.

Hermann Minkowski devised the geometry of space and time.

spacecraft had its own beacon, people on Earth would also see it flashing slower but you would not notice any change.

Geometry of spacetime

Three years later Hermann Minkowski, a German mathematician, found a new way to look at these changes, saying that, thanks to Einstein: "Henceforth space by itself, and time by itself, are doomed to fade away into mere shadows, and only a kind of union of the two will preserve an independent reality." In Minkowski's new geometry of space and time, the space-distances and times we measure really are shadow-like projections of a kind of union of space and time called spacetime. This does make sense of the way we actually experience objects. A soap bubble is a good illustration. Just as we can say that it is an inch (or a centimeter) long, we can say that its existence is 10 seconds long as well. A person might occupy 3 cubic feet in space, and 90 years in time. On the other hand, it's not really possible to have an object that does not last for any time. So, the length of time for which an object lasts is as much a part of it as its length, width, or height.

Lengths and angles

If you look at a yard-long ruler, you rarely see it as one yard long; the length depends on the angle you view it at. The rules of projective geometry (see more, page 76) explain how these changes work. Could it be, Minkowski wondered, that there is an explanation here for why objects shrink according to the speed of the observer?

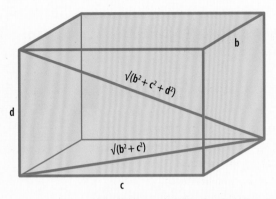

Minkowski's geometry is based on the Pythagoras theorem, seen here in two and three dimensions.

Could that be explained by a new kind of geometry, too? His theory was that every object has a true "extension" or "interval" in spacetime, which we perceive partly as the time it lasts for, and partly as its shape in space. If the object moves rapidly past us (or we move past it), we

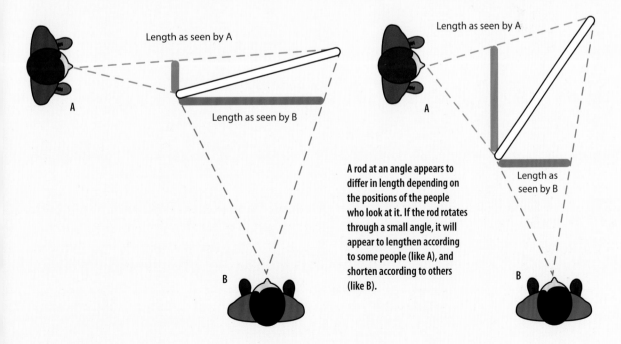

A rod at an angle appears to differ in length depending on the positions of the people who look at it. If the rod rotates through a small angle, it will appear to lengthen according to some people (like A), and shorten according to others (like B).

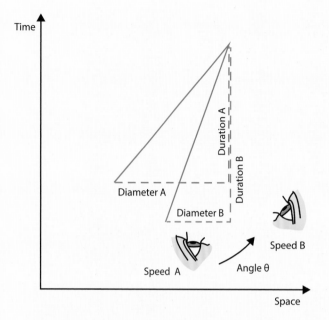

World lines

In the last equation, a is still a kind of length, but it measures both time and space. It is usually called a spacetime interval. The minus sign shows that the time dimension is not quite like a spatial one. This geometrical version of relativity means that the changes of length and duration that fast-moving travelers observe can be explained as a changing point of view. By way of illustration, imagine if you were to walk for 10 minutes and then run for 5 minutes. A plot of the distance you have traveled against time would be like the graph below. The faster you go, the flatter the line becomes. So, the slope of the line indicates speed. We can measure the slope as the angle the blue line makes with the vertical axis. Diagrams like this are used in relativity to plot the paths of moving objects. Usually, they are chosen so that

perceive it as lasting longer in time but being shorter in space. This is like the way a rod that is rotated slightly may appear to lengthen from one point of view, but shorten from another (see the diagram on the previous page). Spacetime is our ordinary three-dimensional space, but with a fourth dimension added to it—the time dimension. Minkowski's geometry is based on one of the oldest and most famous geometrical theorems of all, the Pythagoras theorem:

Two dimensions, $a^2 = b^2 + c^2$
Three dimensions, $a^2 = b^2 + c^2 + d^2$
Four-dimensional spacetime,
$a^2 = b^2 + c^2 + d^2 - (\text{speed of light})^2 \times t^2$

Walking, then running. The horizontal axis is in yards or meters, and the vertical one is in minutes. The walk begins at 10 o'clock. The faster you go the flatter the line becomes.

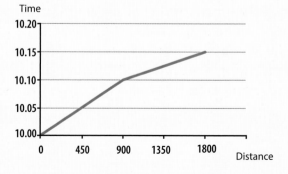

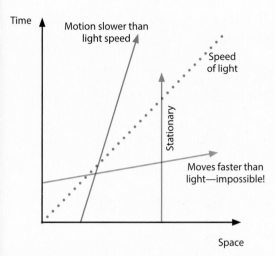

Time

Motion slower than light speed

Speed of light

Stationary

Moves faster than light—impossible!

Space

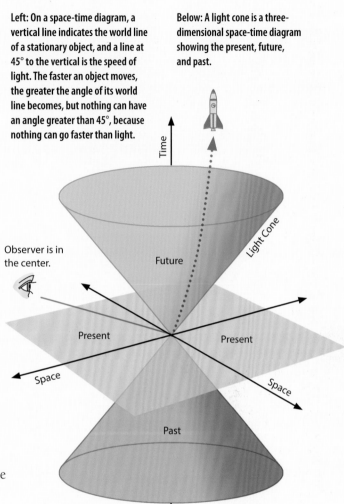

Left: On a space-time diagram, a vertical line indicates the world line of a stationary object, and a line at 45° to the vertical is the speed of light. The faster an object moves, the greater the angle of its world line becomes, but nothing can have an angle greater than 45°, because nothing can go faster than light.

Below: A light cone is a three-dimensional space-time diagram showing the present, future, and past.

Time

Light Cone

Future

Observer is in the center.

Present

Present

Space

Space

Past

the slope of the speed of light is a diagonal. The speed of light is an important element of Minkowski's geometry, because, as Einstein showed, it is a kind of universal speed limit. Nothing in the Universe can go faster than light. Because there are four dimensions in spacetime, a proper map can't be drawn, so instead, just as in the walking-then-running diagram, only one dimension is sometimes plotted. On a space-time diagram, just like the one above, the slope of a line indicates its speed. The paths of objects on the diagram are called world lines.

Light cone

A three-dimensional space-time diagram can be drawn by plotting two dimensions of space against time. The speed of light in this diagram makes a cone, called a light cone. A diagram like this could be drawn for any observer, who is placed at the center, which represents the present position. The lower cone is the observer's past,

and the cone above the observer is their future. Because they cannot exceed the speed of light, they will never be able to move beyond the cone.

SEE ALSO:
▸ Perspective, page 76
▸ Higher Dimensions, page 116

Fractals

EVER SINCE THE ANCIENT GREEKS, SOME MATHEMATICIANS HAVE TRIED TO STUDY THE WORLD AROUND US AS IF IT IS CONSTRUCTED OF SIMPLE GEOMETRICAL SHAPES. This works well for a few things, like crystals and planets. But most real shapes, like islands, trees, and clouds, are not so simple.

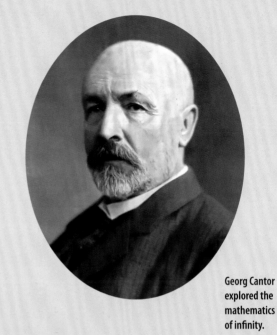

Georg Cantor explored the mathematics of infinity.

Mathematical versions of complicated real shapes like these were found only accidentally, as a result of a new kind of geometry that began in 1870 with the work of German mathematician Georg Cantor. Cantor was the first mathematician to study the concept of infinity properly, and one of his approaches was to apply an infinitely repeated process to the simplest shape of all, a straight line. The process is really simple but produces a complex result: erase the central third of the line, and then erase the central thirds of the parts that remain, and continue that for ever. In the end (whatever that might mean) what is left is mostly empty space, but also an infinite number of infinitely narrow lines, called the Cantor set, or

Making Cantor dust.

Cantor dust. Euclid would have found it impossible to properly define Cantor dust with straight edge and compass, and Descartes could not have found an algebraic formula for it, either. However, a newer geometrical concept made it easier to understand.

Self-similar snowflakes

Like the instruction for the Cantor set itself, the kind of pattern that results is easily described by this novel geometrical concept—self-similarity. An ordinary straight line is self-similar, because zooming in on it will reveal a magnified version which looks similar to an unmagnified version. (In mathematics "similar" means proportionally the same—size is irrelevant.) This is not true of many geometrical shapes. If you zoom in on a curved line, you get a less curved

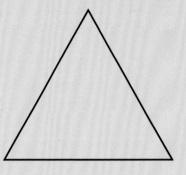

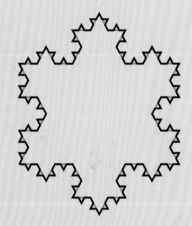

The first four iterations of a Koch snowflake.

line, and if you zoom in on a cube, you don't get a cube. But zooming in on Cantor dust just shows you more Cantor dust, so it is self-similar. Antoine's necklace (see page 149) is a kind of three-dimensional version of the Cantor set.

Making mathematical snow

In 1905, Helge von Koch, a Swedish mathematician, applied Cantor's approach to define a self-similar shape now called a Koch

Helge von Koch.

curve, which is usually shown as part of a shape called the Koch snowflake. The snowflake starts as a triangle. The middle third of each of its sides is replaced by a triangle, and this process is repeated ("iterated") endlessly. The result is another kind of shape which would have baffled earlier geometers, since it has an infinitely long edge but a finite area (which is 8/5 the area of the initial triangle).

Defining dimensions

Self-similarity is all around us. Clouds, ferns, snowflakes, the blood vessels of the body, the way the wind blows, the path of a lightning strike, many other natural phenomena, and the stock market are self-similar. On its own, the idea of self-similarity is not a very mathematical one, but in 1918 German mathematician and poet

Felix Hausdorff found a way to pin it down, by introducing the concept of a fractional dimension. This concept makes little sense if we think of dimensions in terms of ordinary shapes. A point is zero-dimensional, a line, which is defined only by its length, is one-dimensional. A hexagon is two-dimensional, because its flat shape has only height and width. It's hard to imagine how to fit a half-dimensional shape into this series, or a 4/3-dimensional one. But how many dimensions does a Cantor set have? It begins with an ordinary straight line, which is one-dimensional, but it ends with points, which are zero-dimensional. So we could say the whole set has a fractional dimension (a Hausdorff dimension), somewhere between 0 and 1. How can we calculate an actual value for the Hausdorff dimension of a Cantor set? The idea of calculating a value for any dimension

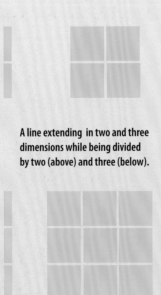

A line extending in two and three dimensions while being divided by two (above) and three (below).

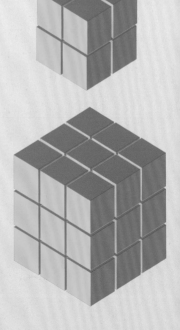

Above: A three-dimensional version of Cantor dust.

LOGARITHMS

We could find the value of n in $2=3^n$ in several ways, but the simplest method uses logarithms. We can raise a number to a power, whether that power is a whole number or a fraction:

$$3^2 = 9, \ 3^3 = 27, \ 3^{0.63} \approx 2$$

But what if we want to go the other way? So, instead of getting from 3 to 9 or 3 to 27, we want to find a mathematical way of going from 9 or from 27 to a power of 3. From $3^2 = 9$, we can see that "the-power-to-which-we-have-to-raise-3-to-get-9" is 2. A shorter version of "the-power-to-which-we-have-to-raise-3-to-get-9" is "logarithm to base 3", or even more briefly, $\log_3$. So, $\log_3$ of 9 is 2, $\log_3$ of 27 is 3 and $\log_3$ of 2 is about 0.63. Logarithms are fiddly to calculate by hand, but many calculators will work them out.

this becomes $2 = 2^1$, for the square, $4 = 2^2$, and for the cube, $8 = 2^3$, so the formula seems to work. A similar formula works if we make more cuts. If we make 3 cuts per dimension, we get 3 small lines from a line, 9 small squares from a square, and 27 small cubes from a cube. So this time the equation becomes: number of smaller versions = $3^{\text{dimension number}}$. And, sure enough, for the line, $3 = 3^1$, for the square $9 = 3^2$, and for the cube $27 = 3^3$. So now we can write a more general formula for the number of dimensions: number of smaller versions = number of cuts per dimension$^{\text{dimension number}}$.

seems odd, because for familiar shapes like lines, squares, and cubes, the number of dimensions is obvious. But we can find a way to do this.

A formula for dimensions

If we divide a line by 2, we get two smaller lines. If we divide each of the two dimensions of a square by 2, we get 4 smaller squares. And if we divide each of the three dimensions of a cube by 2, we get 8 smaller cubes. We can summarize these results in a formula: number of smaller versions = $2^{\text{dimension number}}$. For the line,

The first five iterations of the Koch curve.

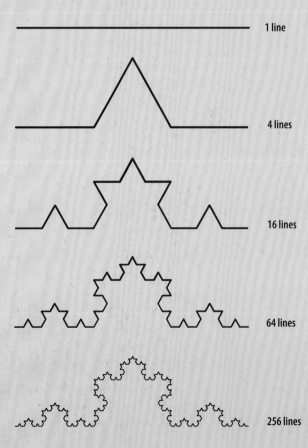

1 line

4 lines

16 lines

64 lines

256 lines

Fractional dimensions

We can use this formula to work out the value of the Hausdorff dimension of a Cantor set. Dividing a Cantor set into 3 gives 2 smaller Cantor sets (plus one blank area). So our formula becomes $2 = 3^{\text{Hausdorff dimension}}$. This gives the value of the Hausdorff dimension as 0.63, because $2 \approx 3^{0.63}$ (see logarithm box, previous page, for how this is calculated). We can use the same formula as before to work out the dimension of the Koch snowflake: as the figure shows, each time we cut a line in the Koch curve into 3, we get 4 smaller lines. Putting this into the formula number of smaller versions = number of cuts per dimension$^{\text{Hausdorff dimension}}$ gives $4 = 3^{\text{Hausdorff dimension}}$. And, $3^{1.26} \approx 4$, so the Hausdorff dimension is about 1.26.

Real-world values

The Hausdorff dimensions of naturally self-similar shapes can be calculated, too, (though actually working out precise values can be very difficult). A typical cloud, for example, has a Hausdorff dimension of about 1.35, and the coastline of the island of Great Britain is about 1.26.

Complexity from simplicity

The idea of defining complicated shapes by iterating simple instructions can be applied to equations, too. In the 1910s, French mathematicians Gaston Julia and Pierre Fatou studied such equations when they were investigating the patterns that can be formed on a two-dimensional surface. An equation like $y = x^2 + a$ can be iterated like this: we start by assigning x and a any values, like $x = 1$ and $a = -0.1$, and then calculate y.

$$y = 1^2 - 0.1 = 0.9$$

Now we set x to this value of 0.9, and calculate again.

$$y = 0.9^2 - 0.1 = 0.71.$$

We repeat this process several times, each time setting x to the previous answer, and we get 0.9, 0.71, 0.4041, 0.06329681, −0.095993514, −0.090785245,

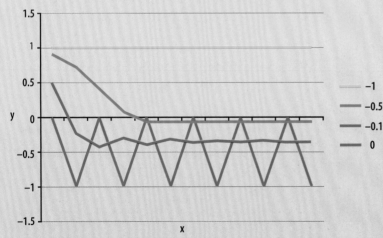

The iterated function $y = x^2 + a$ behaves differently for each value of a (listed on the right of the diagram). If a is greater than zero, the function goes to infinity.

A Julia set as rendered by a computer.

Auch von dem Falle, daß ℌ sowohl „verkettete" als „unverkettete" Punkte enthält, können wir uns ein Bild machen, indem wir die Konstruktion von oben nochmals so ausführen, daß die Dreiecke sich nicht mehr berühren, d. h., anschaulich gesprochen, indem wir unser Schema passend „auseinanderziehen", wie dies in nebenstehender Figur 3 angedeutet ist. Dann ist das entsprechende 𝔅" von unendlich hohem Zusammenhang. Zwei Punkte von ℜ sind offenbar dann und nur dann

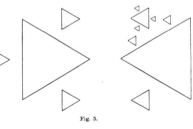

Fig. 3.

verkettet, wenn sie dem gleichen Dreieck angehören. In jedem noch so kleinen Bereich D, der Punkte von ℜ enthält, liegen dann unend-

Left and below: The most complicated plot that Julia and Fatou would have seen was this one, in a paper published in 1925.

208   HUBERT CREMER:

Wir gehen von zwei gleichseitigen Dreiecken △ $A_1 A_2 A_3$ und △ $A_1 A_4 A_5$ mit der Seite a aus, die an der Ecke A_1 aneinanderstoßen (Fig. 2). Sie bilden zusammen den geschlossenen polygonalen Zug $p_1 = A_1 A_2 A_3 A_4 A_5$, der die Ebene in 3 Bereiche teilt:

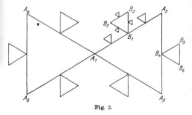

Fig. 2.

1. Das Innere von △ $A_1 A_2 A_3$: $\mathfrak{B}_1$.

2. Das Innere von △ $A_1 A_4 A_5$: $\mathfrak{B}_1'$.

3. Den Bereich $\mathfrak{B}_1''$, der den unendlich fernen Punkt enthält und vom ganzen polygonalen Zug p_1 begrenzt wird.

In die Mitte jeder der Seiten von p_1 setzen wir die Spitze eines gleichseitigen Dreiecks von der Seitenlänge $\frac{1}{3} a$[1]), dessen Seiten den Seiten von p_1 parallel sind, und das (d. h. dessen Inneres) ganz in $\mathfrak{B}_1''$ liegt. △ $B_1 B_2 B_3$, △ $B_4 B_5 B_6$ sind zwei dieser Dreiecke. p_1 bildet mit den so konstruierten Dreiecken einen geschlossenen polygonalen Zug p_2, dessen Ecken in der Reihenfolge $A_1 B_1 B_2 B_3 B_1 A_2 B_5 B_6 B_4 A_3$ aufeinander folgen. p_2 teilt die Ebene in $3 + 6 = 9$ Bereiche, nämlich $\mathfrak{B}_1$, $\mathfrak{B}_1'$, das Innere der sechs oben konstruierten Dreiecke und den vom ganzen Zug p_2

−0.091758039, −0.091580462, −0.091613019, and so on. The answer is settling down to about −0.09161. Depending on our choice of a, one of five things happens next. The answer may remain at a fixed value; move steadily to a fixed value; oscillate endlessly; oscillate at first, but gradually settle down to a fixed value, or go to infinity.

Using complex numbers

This is already quite a complicated outcome from a simple equation, but things get more interesting if the same idea is applied to a two-dimensional plot. To do this properly involves complex numbers. (Complex numbers combine real numbers like 1, 2, and 3 with an imaginary unit i to create a more complete set of values for answering tricky algebra problems.) This is how it works: we start with any values of x and y we like, such as x = −2 and y = 0. We are going to treat this pair as a coordinate and draw a dot there, but first we work out what color it should be. To do this, we select two more values, a and b. Then we write $(X + Y)^2 = (x + y)^2 + (a + b)$. We solve this to calculate X and Y, then feed this answer back into the equation, setting x and y to the values of X and Y we have calculated, and calculating new values. We continue to iterate and see what is happening to (X + Y). For some values of a, b, x, and y, (X + Y) will remain the same, in others it goes rapidly to infinity, and so on. Now, we assign a color for each kind of answer. One color for steady values, another for values that go slowly to infinity, another if values go rapidly to infinity, and so on. Then, we color our dot at (−2, 0). We do this for many other dots and finally we will get a pattern called a Julia set (see previous page).

The Julia set is self-similar. Each of the two large shapes is surrounded by smaller versions of the same shape, which are themselves surrounded by even smaller ones, and so on for ever.

Waiting for the computer

Sadly, Gaston Julia never actually saw beautiful plots like the one on the previous page; without electronic computers equipped with graphics facilities, such plots would take thousands of hours to make. So it's not surprising that these self-similar patterns were not very well known until computer graphics

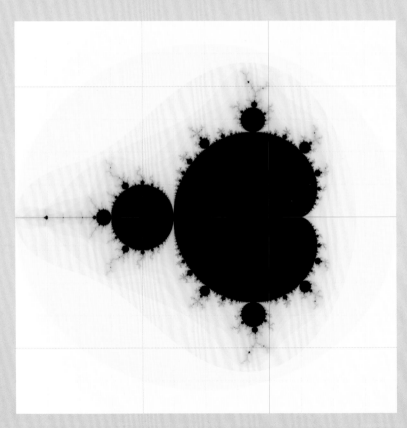

The Mandelbrot set is the most famous of all fractals.

A computer-generated landscape based on fractals looks real because many natural shapes are fractals already.

became readily available, starting in earnest in the 1970s. It was only in 1975 that these shapes were even given a name: fractal. That name refers to the shape's fractured or fractional dimensions. This term was suggested by Benoit Mandelbrot, whose name is given to an even more complex and beautiful self-similar pattern. Mandelbrot sets are produced in a very similar way to Julia sets, except that in an expression like $(x + y)^2 + a$ + b, it is a and b which are iterated, rather than x and y. Graphics software makes a great deal of use of fractal images, which can be designed to look like forests, galaxies, rivers, and many other complicated natural things.

SEE ALSO:
▸ Euclid's Revolution, page 44
▸ Higher Dimensions, page 116

Unknown Geometry

René Descartes redefined geometry and unleashed a wave of discovery.

GEOMETRY HAS PLENTY OF PROBLEMS STILL UNSOLVED, and, unlike the remaining problems in other areas of mathematics, many of them are easy to understand. You don't necessarily have to be a world-class mathematician to solve them, either. Often, seeing the solution to a geometrical problem involves having a vivid imagination. This is especially true of the many unsolved geometrical problems which concern the properties of higher-dimensional objects or spaces.

Now that the Poincaré conjecture has been solved, the most significant challenge in geometry today is probably the search for a true knot invariant (see more, page 146). Meanwhile, mathematicians all over the world are seeking to link arithmetic and geometry in new ways (see box, opposite).

Buried secrets

Many breakthroughs in geometry are not a result of trying to solve a known problem. Often, they are made by mathematicians striking out into new areas, or ones that have been neglected. This can provide new ways to solve old problems easily. So, René Descartes made huge strides in analytic geometry (see more, page 98), Peter Guthrie Tait in the mathematics of knots (see more, page 146), and Benoit Mandelbrot in fractals (see more, page 166). However, even mathematical fields which have been carefully sifted and studied for centuries sometimes contain hidden treasures. More than 2,000 years since the ancient Greeks had made very thorough studies of how polygons could be drawn using a straight edge and compass (see more, page 64), Carl Gauss found that a 17-sided polygon could be constructed in that way, and was so pleased with his result that he asked for it to be inscribed on his tombstone. (The stonemason refused, saying it looked too much like a circle.)

Several other theorems have been found long after they could have been. Who knows how many more discoveries like these could be easily found with just a little well-directed puzzling?

Mystery drive

Why should anyone spend time trying to explore uncharted areas of geometry, or tackling ancient geometrical puzzles? Success is highly uncertain, and the solutions are unlikely to make anyone rich or famous. One reason is the

At the age of just 19, Carl Gauss did what several generations of ancient Greek geometers failed to do by drawing a 17-sided heptadecagon with a straight edge and compass.

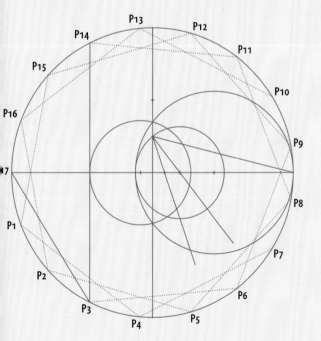

LANGLANDS PROGRAM

In the 16th century, geometry leapt forward thanks to Descartes and Fermat joining algebra and geometry (see more, page 98). The Langlands program was launched by American mathematician Robert Langlands (below) to do something similar for geometry and an area of arithmetic called algebraic number theory. The program began in January 1967 with a letter from Langlands to André Weil, one of the greatest mathematicians of the day. Langlands suggested a very powerful—but very tentative—link between number theory and geometry. Humbly, he wrote, "If you are willing to read it as pure speculation I would appreciate that." He continued: "If not—I am sure you have a waste basket handy." It's just as well Weil didn't throw the letter away. Today, the Langlands program has become the largest project in mathematics, and one of the most important.

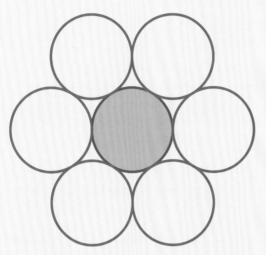

In two dimensions, the kissing number is 6, which is simple to see. In three dimensions it is 12, which is not simple at all to see or figure out.

same thing that drives people to solve riddles, watch mystery films, do crosswords, or read detective stories. The fun comes from untangling something complex. Another is the impulse that sends explorers to remote mountains or the Moon: to reach a place that no one has visited before. However, new discoveries in mathematics are, at least in one way, more exciting than conquering new peaks or visiting other planets. A mathematical discovery is a fact about the Universe that is true everywhere, and always will be. As Albert Einstein once said, "Equations are for eternity."

The Kissing Problem

The kissing number for a sphere is the number of other spheres that can touch it at the same time. For spheres in a row (that is, in one dimension), the kissing number is 2. For spheres on a surface (two dimensions), it is 6. Given that those answers seem so easy, it's surprising how tricky the question gets for higher dimensions. In three dimensions, 12 spheres can fit around another, but there is a lot of spare space left, almost enough to fit a thirteenth sphere. But not quite! So the kissing number is 12. Kepler (see more, page 92) seems to have been the

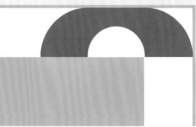

MORLEY'S MIRACLE

Morley's theorem could have easily been found by the ancient Greeks, or by anyone else playing with a ruler over the next thousand years or so. Because it wasn't, and because it's such a startling theorem, it is called Morley's Miracle after American mathematician Frank Morley, who finally discovered it in 1899. Just take any triangle and trisect each of its angles (though the Greeks could not do this with straight edge and compass, they knew many other ways to manage it). Draw lines through these trisected angles, until they meet. This will always produce an equilateral triangle, no matter what shape the original triangle was.

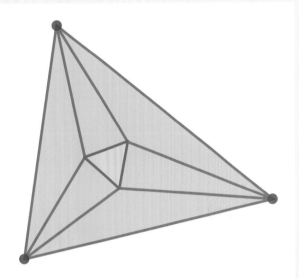

first to state this, in 1611, but it was not proved mathematically until 1953, even though it's easy enough to demonstrate with actual spheres. In four dimensions, the kissing number (which is 24) was only discovered in 2003, and for most higher dimensions it is still unknown.

The Sofa Problem

Imagine trying to maneuver a large sofa (or couch) around a corner in a corridor. Obviously, the bigger the sofa, the tougher the task, but

what is the "sofa constant," the largest sofa, of any shape you like, that you could possibly fit? For a real sofa, you could try angling it up at one corner, but this makes the problem much more complicated, so instead let's say that the sofa must remain level. If you can't work out the answer, you're not alone—no mathematician has ever solved the problem. All that has been done

Moving a sofa is a difficult problem in practice but a complete mystery in mathematics.

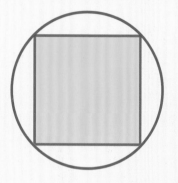

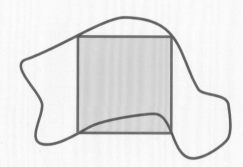

A square fits with this warped version of a circle. Is that true in every case?

is to narrow it down a little: for a corridor 1 yard across, the "sofa constant" lies between 2.2195 and 2.8284 yards.

The Inscribed Square Problem

If you draw a circle, it is easy to see that you can draw a square in it, so that each of its corners lies on the circle. This is called an inscribed square. If you distort the circle (without allowing it to cross itself), it is usually easy to find somewhere on it where you can inscribe a square. However, no one has yet managed to prove that every such shape has an inscribed square.

The Happy-Ending Problem

Draw five dots on a piece of paper, the only rule is that you must not line them up.

The red dots would make a convex pentagon but the green and blue ones would not.

You should be able to draw a quadrilateral, using four of the points as its corners. But you won't necessarily be able to draw a convex pentagon (where the corners have inner angles smaller than their outer ones.) To be sure of doing that no matter how the dots are scattered, you need nine of them. To be sure of being able to draw a hexagon using randomly scattered dots, 17 dots are needed. To be sure of drawing a heptagon

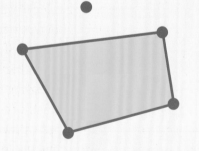

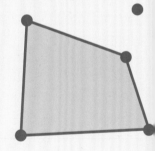

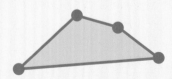

The cube has 11 possible nets, shown here, and it seems likely that all possible polyhedra have at least one. However, no one can say for sure.

(a seven-sided figure) … no one knows how many dots are needed. People think 33 is enough, but no one has proved it. This is called the Happy Ending Problem because George Szekeres and Esther Klein enjoyed working together on it so much, they ended up getting married.

Ulam packing conjecture

Some shapes, like cubes, will fit perfectly into a box, leaving no space unfilled. Others, like cylinders, fit quite well but leave a bit of space. Spheres are worse than cylinders; they leave a lot of space. Is anything worse than spheres? No one can say for sure. Stanislaw Ulam, whose conjecture this is, was a Polish nuclear physicist, who helped build the first nuclear reactor and to design nuclear spacecraft to travel to other stars.

Dürer's conjecture

Albrecht Dürer, who invented nets (see more, page 37), may possibly have wondered whether every polyhedron has a net. However, this conjecture was actually not recorded until the 1970s, presumably because it sounds so likely to

be true. Which makes it all the more surprising that nobody has ever managed to prove it (assuming we rule out the polyhedra with dips in them, called concave polyhedra).

Still lost in the forest

In 1956 Richard Bellman, an American mathematician and computer engineer, asked: if you find yourself lost in a forest of which you know the shape, what is the best way out? It might seem that walking in a straight line would be best, but what if you are not far from the forest's edge and set off parallel to that edge? In that case, a zig-zag route would be much quicker. More than half a century later, this question has only been answered for a rectangular forest, a circular forest and a few other shapes. For triangular forests and for most other shapes, the solution has still not been found.

SEE ALSO:
▶ Perspective, page 76
▶ Higher Dimensions, page 116

Glossary

Absolute geometry

A type of geometry in which none of the theorems rely on the parallel postulate, nor any of its alternatives. See Euclidean geometry.

Algebraic geometry

Geometrical methods used to solve equations.

Analytic geometry

Geometry in which shapes and curves can be plotted using coordinates.

Axioms

Basic assumptions on which theorems are based. An example is: "any two points can be joined by a straight line." Different kinds of geometry can be based on different sets of axioms.

Complex number

A number with a real part and an imaginary part, such as 7 + 2i, where i= $\sqrt{(-1)}$.

Conjecture

A statement of theory believed to be true, but not proved.

Differential geometry

Differential calculus, or differentiation, is a mathematical tool for dealing with things that change. In differential geometry, differentiation and related techniques are used to study curves and surfaces.

Elliptical geometry

An alternative to Euclidean geometry in which the parallel postulate is not used, the angles in a triangle add up to more than 180°, and there are no parallel lines.

Euclidean geometry

Geometry which includes the parallel postulate, based on the work of the ancient Greek mathematician Euclid.

Hyperbolic geometry

An alternative to Euclidean geometry in which the parallel postulate is not used and the angles in a triangle add up to less than 180°.

Hypersphere

A four-dimensional version of a sphere.

Imaginary number

The square root of -1, or a multiple of it.

Non-Euclidean geometry

Any kind of geometry that does not use the parallel postulate.

Parallel postulate

The statement that every line has another line parallel to it, which is to say that the two lines are a constant distance apart everywhere.

Prism

A three-dimensional shape in which two parallel sides are identical polygons, one directly above the other.

Projective geometry

The type of geometry that explains the way that three-dimensional objects are drawn (or projected) on flat surfaces.

Radian

A measure of angle, equal to $180°/\pi$, which is about $57.3°$. If a distance is marked off along a circle which is equal in length to the radius of that circle, and lines are drawn from the ends of that distance to meet at the centre of the circle, then the angle between those lines is defined as 1 radian.

Solid

In geometry, a solid is a three-dimensional shape, such as a sphere or cube.

Spherical geometry

The kind of geometry which applies to shapes on the surface of a sphere.

Straight edge and compass

The only tools which many ancient Greeks relied on to study geometry. A straight edge is like a ruler with no marks on it, and a compass is what we would call dividers.

Theorem

A mathematical statement which has been proved to be true.

Topology

An area of geometry studying certain properties of shapes that remain the same even when the shapes are distorted. Those properties include the number of holes in the shape, but do not include lengths or angles. This is the most active area of research and development in geometry today.

Trigonometry

The area of geometry which deals with triangles.

Index

Cataloging-in-Publication Data has been applied for and may be obtained from the Library of Congress.

ISBN 978-1-62795-138-8

Design: Bradbury & Williams
Copy editor: Meredith MacArdle
Proofreader: Julia Adams
Consultant: Kevin Adams
Picture Research: Clare Newman
Cover Design: Wildpixel Ltd

Publisher's Note: While every effort has been made to ensure that the information herein is complete and accurate, the publisher and authors make no representations or warranties either expressed or implied of any kind with respect to this book to the reader. Neither the authors nor the publisher shall be liable or responsible for any damage, loss, or expense of any kind arising out of information contained in this book. The thoughts or opinions expressed in this book represent the personal views of the authors and not necessarily those of the publisher. Further, the publisher takes no responsibility for third party websites or their content.

SHELTER HARBOR PRESS

603 West 115th Street Suite 163
New York, New York 10025

For sales, please contact
info@shelterharborpress.com

Printed in China.

10 9 8 7 6 5 4 3 2 1

PICTURE CREDITS

Alamy: Chronical 142r, David J Green 145c, Historic Collection 121c, 126, History & Art Collection 144t, ITAR-TASS News Agency 159t, Science History Images 4-5, 131, 151, The Picture Art Collection 92bl, The Print Collector 110t; **Getty Images:** Corbis Historical, Hulton Archive 42t; **NASA:** 19t, 43; **Public Domain:** 39, 136, 145tl; **Shutterstock:** Alhovik 109t, Mila Alkovska 64, Serge Aubert 18, Robert Biedermann 85bc, Dan Breckwoldt 12b, Cyrsiam 76bl, Darq 78, Dotted Yeti, 125b, Peter Hermes Furian 129bl, 129br, Aidan Gilchist 144cl, Itynal 94, Victor Kiev 113b, Mary416 96t, Morphart Creation 52, 53b, 86, 112, 133, Nice Media Production 30, Hein Nouwns 125t, Onalizaoo 27 85br, Pike Picture 124t, Saran Poroong 29, Potapov 160, Robuart 146, Andrew Roland 142tl, Slava 2009 67t, Joseph Sohm 26, Spiroview Inc 144cl, Taiga 73bl, Zoltan Tarlacz 106, Vchal 8, Ventura, 50, Vex Worldwide 75br, xpixel 110b; **The Wellcome Library, London:** 14tr, 57, 101, **Wikimedia Commons:** 6t, 7b, 9, 10l, 10r, 11br, 13, 14b, 16t, 19b, 20br, 21, 22b, 24l, 25tl, 25cr, 34, 37tr, 38t, 42b, 44, 47, 53tl, 55b, 59, 63, 66tr, 66cl, 67b, 68br, 70, 71, 72b, 73cl, 73br, 74, 75l, 76tr, 77, 79, 81, 82-83b, 83r, 85t, 88l, 88r, 89, 90, 91t, 91b, 92tr, 93, 95, 96, 96b, 98b, 105, 107tr, 107bl, 108, 109b, 113t, 114, 116, 117t, 117b, 120, 122, 130, 137, 138, 139, 140, 141b, 147, 148, 149, 150, 153, 155t, 156tl, 156b, 161, 162b, 166, 167, 168bl, 169, 171, 172, 173, 174, 175bl, 175br, NASA 162t; **Roy Williams:** 16-17b.

ILLUSTRATIONS

Shutterstock: NoPainNoGain 87t.
Wikimedia Commons: Creative Commons Attribution-Share Alike 2.0 Austria license/R J Hall 20br, Creative Commons Attribution-Share Alike 3.0 Unported license/Win 21, Creative Commons Attribution-Share Alike 3.0 Unported license/Pbroks13 at English Wikipedia 22t, Creative Commons Attribution-Share Alike 2.5 Generic license 23bl, Robert Webb's Stella software http://www.software3d.com/Stella.php 36b, Creative Commons Attribution-Share Alike 4.0 International license/Krass 60b, Creative Commons Attribution-Share Alike 4.0 International license/EdPeggLr 71, Creative Commons Attribution-Share Alike 3.0 Unported license/Geometry guy at English Wikipedia 74tl, Creative Commons Attribution-Share Alike 2.5 Generic license/McSush 87c, Creative Commons Attribution-Share Alike 3.0 Unported license/Dr Fiedorowicz 121t, Creative Commons Attribution-Share Alike 3.0 Unported license/Alecmconroy at the English language Wikipedia 130b, Creative Commons Attribution-Share Alike 3.0 Unported license/Pbroks13 134b, Creative Commons Attribution-Share Alike 3.0 Unported license/Brilee 89 138l, Creative Commons Attribution-Share Alike 3.0 Unported, 2.5 Generic, 2.0 Generic and 1.0 Generic license/Napy1Kenobi 138r, Creative Commons Attribution-Share Alike 3.0 Unported, 2.5 Generic, 2.0 Generic and 1.0 Generic license/Frederic Michel 144t, Creative Commons Attribution-Share Alike 3.0 Unported license 166, Creative Commons Attribution-Share Alike 3.0 Unported license, Robertwb 176tr, Creative Commons Attribution-Share Alike 3.0 Unported license/Claudio Rocchini 176b, Creative Commons Attribution-Share Alike 4.0 International, 3.0 Unported, 2.5 Generic, 2.0 Generic and 1.0 Generic license/R. A. Nonenmacher 179.

Publisher's note: Every effort has been made to trace copyright holders and seek permission to use illustrative material. The publishers wish to apologize for any inadvertent errors or omissions and would be glad to rectify these in future editions.